Fast and Flawless Pricing

A guide to pricing and business for decorators

2019

First published in the United Kingdom in 2019

22-12-2019

5 X 8

Thanks to my wife Tracey for being forever by my side and giving me support.

Thanks to Ian and Lyndsey at PaintTech Decorators for showing me many things that have opened my eyes to how decorating could be.

Thanks to Andrew at Alizeti for giving me an insight into the world of a multimillion pound decorating company.

Last but of course, not least thanks to Mum and Dad for being there when I need them and teaching me to look at the world in my own way.

Contents

Preface

Years ago, when I was at the age of 22, I decided to leave my decorating firm and work for myself. It will be easy I thought. What I didn't realise at the time is that my boss did a lot of stuff behind the scenes, all I had to do was turn up and decorate.

Pricing had always been a challenge for me, I didn't really know how to go about it, so I asked a few of my self-employed decorator friends. They were reluctant to give me too much information because I was a competitor but also, I have now realised, they didn't really know themselves.

Of all the skills I had learnt over my apprenticeship, pricing was not one of them. While we all think it's important to be able to cut in or hang wallpaper, really the most important skill once you have your own business is pricing. It is a skill that needs to be learnt and developed over time just like all your other skills. This skill however really does pay well, if you are good at it then you will earn a good living and have a work life balance that suits you.

I was lucky enough at 27 to get a job teaching Painting and Decorating at a local college. More recently I have gone part time and gone back on the tools. I have also set up a private training organisation in conjunction with partners in London to deliver training to the trade.

These experiences mean that I have taught and spoke to many decorators all over the country. Many of the decorators who do the airless spraying course chat about their businesses, and what has become clear to me is that pricing jobs is a challenge for most people. It has been a relief to find out it was not just me.

I believe that our "lot" as a decorator can be improved a great deal by changing a few things that we do. The first thing I believe we should change is to be more productive, and that is the topic of my first book "Fast and flawless" which is a guide to airless spraying.

Once we are more productive, the next thing we need to change is our pricing system. It's no good doing the work 4 times faster if you charge £120 a day because the only person that benefits then is the customer.

The main aim of this book is to discuss pricing and the various approaches that you can take. I get quite a few messages and emails that are along the lines of, how much would you price for this job? There is usually a picture with the message. Unless it's a friend I try to sidestep these types of question.

Why is this?

Well there is no upside for me. For example, if I say the job should be priced at £500. They then either get the job or they don't. If they get the job and they don't make money, then it's my fault. If they don't get the job, then it's my fault. If they get the job and make money, then I am forgotten about. Two out of the 3 outcomes casts me as the bad guy. These are not odds that I want to take.

Also, there are many variables. I always find it interesting when people start discussing "day rate" I find that different numbers are bandied about. Dave charges £80 a day, Harry charges £130, Bill charges £200.

These figures mean nothing at all.

What if Dave (the cheap £80 a day one) is useless and takes all week to do what Bill can do in a morning. The job would cost £400 if Dave did it, and £100 if Bill did it. I know you are thinking that you would never get such a big difference in people's output, but you do.

Where you are based geographically is also a factor. If you are in an area with loads of work and not much competition then you can get higher prices, if you are in a city with lots of competition then maybe you will have to be keener with your price.

Where you are personally is also a factor. You may be at the start of your career and you want to build up a business that will pay you a full-time wage and pay the mortgage. Your prices need to get the work in this case.

If you are older and your mortgage is paid off, then you are under less pressure financially so you can afford to gamble a little and put in higher prices and work less.

You get the picture, your personal work rate, the area that you live in and your financial position are all unique to you, so you need to understand yourself how to price, and then do it.

This book is designed to show you the way to understand pricing and your own business and explain how you can go about pricing your own jobs. I will also give you pointers on where to go if you need further advice and help you develop your pricing skills.

It is only an introduction to get you going, I am not going to get bogged down with too much complexity, maybe save that for another book, for now I just want to convince you to change your approach and give you a few tools so that you can do that.

I am going to use figures plucked out of thin air to illustrate my calculations, real rates are constantly changing, and you may be reading this book five years after I have written it so the figures would be out of date. I will however show you how to get your own rates and throw in the odd real figure that I know people are using.

This book will also look at topics that are related to pricing such as negotiation and positioning yourself in the market, I want to discuss these things because I feel that they are important things to understand if you are going to build a decorating business, and although the price is very important it is not the only factor.

The way I have written the book is slightly different to my last one. In some of the chapters I am going to tell a story of a young decorator who makes his way through his professional life. It will chart his ups and downs and the mistakes he makes along the way. In between the story chapters I am going to weave in the lessons that need to be learned. These chapters will be factual "how to" chapters. This way it will be easier for you to relate to. In fairness most of the book is factual chapters, the story is just a bit of fun to make the book a bit more interesting.

You could skip the story bits and still learn how to price but I advise that you don't do this because the story contains a lot of the pitfalls that decorators stumble across while trying to grow their business.

I am by no means an "expert" I am not a quantity surveyor or a professional estimator, I am a decorator who has learned the hard way, and is still learning. When developing a pricing strategy for your own business I recommend that you speak to someone local who does it for a living and spend some time with them to develop your own rates.

Finally, I hope you enjoy the read, I am going to keep it lively and chatty and try and keep you interested.

Chapter 1 – Other trades

I am going to start by talking about other trades, now I know you are not really interested in what the joiners or plasterers do. Floor layers and electricians either. But on reflection you are a bit sick that they seem to earn more than you do.

I want to start by having the tough discussion about where we are in the pecking order as decorators. This is not just where other trades see us, but it's also how the public see us and how in a lot of ways, we see ourselves.

One of the things that I am trying to do is change this perception of our trade.

Before we discuss this, I want you to know that this is not how I see our trade, and later on I will set out how I actually see things and how we can go about changing the perception that people have of us.

I used to work in a construction department of a local college. All the trades are in one big staff room, there are plumbers, electricians, joiners, bricklayers and plasterers. Oh and of course the painters. This is what they call us too – painters, we are not decorators or painters and decorators or decorative artists.

Why is this?

Well once upon a time decorating used to be based in the art department and then someone decided that we actually didn't belong there, but we belonged in the construction department. You see when they build houses, all the trades are there including decorators.

We moved into the construction department. The syllabus changed too so that a lot of things that decorators used to study went out of the window and the focus changed. Believe it or not they even tried to take wallpapering out of the course!

The other trades only ever see us emulsioning walls and glossing skirtings on site. They actually think that's all we do. I know this is a generalisation, but you know what I mean.

Anyway, back to the main plot. We have all the trades in the staff room. A new chap started in electrical. He came over chatting and pointed out that he or at least his trade was the top of the building food chain.

Now I saw red at this point but once I had calmed down, I reflected on this. Of course, there is a hierarchy of the trades. It goes something like this; -

Electricians
Plumbers
Joiners
Bricklayers
Plasterers
Painters

Why is this?

Well they are ordered by the amount of money they earn because as a society we value people who earn more money. Now I know that you will get variations and that some joiners will earn more than plumbers and some brickies will earn more than joiners but there you have it.

Why are we at the bottom of the pile? If you look at day rates for all the trades, then our day rate is the lowest. If you look at some of the work that painters turn out on site, it is very poor (I know we are on a price and we have to make it pay) and I think that some decorators even run themselves down.

Then of course there is the “if you can piss you can paint” that is thrown in your face. It’s just “banter” lads, we are told.

Let’s just think about that for a second. What they are basically saying is that anyone can do what we can do, and no-one can do what they do. Where has this come from?

I think it has come from 2 places, first of all your mum or dad probably did the decorating at home, they could not afford a decorator, so they did it themselves. However not many people rewire their own house, there are regulations in place to say that you can’t.

Secondly some professional decorators turn out very poor work that people look at and think that they could do better themselves.

I am interested in boats and I have built a Dutch barge from scratch. Except for the welding I did all the work myself. I did the joinery work, fit the windows, did the electrical work, did the plumbing work and of course I did the painting. All boaters are like me, they have to be a jack of all trades otherwise the hobby would become too expensive. Wiring is not difficult, and neither is plumbing pipes in.

Painting however is very difficult to do well. To get a professional finish on your boat you need to get someone in, no boater that I know can do a good job themselves unless they are a painter or sprayer.

To paint or decorate something really well takes a high level of skill that cannot be acquired by looking on YouTube. It takes years of honing your skill.

This is good news for us.

Unfortunately, a lot of decorators have got a little lazy and let their skill level drop to quite a low standard. Decorators that can produce jaw dropping finishes on woodwork and that can hang expensive wallpaper to a high standard are few and far between.

Why am I going on about this stuff when this book is supposed to be about pricing? The reason is that first of all before you start pricing you need to know your own ***value*** to the market.

This value can be changed by you.

The money that you can charge is directly related to the value that you give the customer. Value is what you need to focus on. When I say value, I don't mean cheap, I mean the value of what you produce. So, for example if I spray a garage door then that adds value to the customers house and to their life by having a nice garage door. It is important to know what the value of what you do is. We will come back to this again and again.

Where do I fit into the hierarchy in the staff room? Well I can tell you that I am not at the bottom. I have decided to place myself where I want to be, I will let you decide where you think that is. I obviously keep this fact to myself, no point in upsetting everyone's view of the world.

Why have I told you this?

I have not told you to boast about income, that's not my style, I have told you because I am just like you. I was a CITB apprentice working for a big painting firm. I have changed a few things about the way that I work and now I earn a good living.

You could be thinking one of three things here.

1. Well Pete, I don't believe you.
2. I believe you but I don't think I could do that.
3. I believe you and I am going to do the same.

I hope you are thinking number three otherwise you may as well put the book down and get on with your life. If you do want to move forward with your business, then stick around and take on board what I have to say.

Chapter 2 – Going it alone

To make the subject of pricing a little easier to discuss I have decided to tell the story of a fictitious decorator. I am going to call our decorator Richard. This character is completely made up and is not based on any decorators that I know. I am going to look at Richard's career through his life right from the start.

Richard was quite good at school, not a real highflyer but not in the bottom set either. He quite enjoyed the day to day structure that school gave him. The years in school flew by and before long it was time to leave and decide what the next step was.

Richard had no idea what his next step was, he was quite good at art but also quite good at making things. A local decorating firm was looking for an apprentice and Richard thought that this sounded good. The decorating company did all kinds of work, from decorating large commercial buildings to small domestic jobs such as decorating a lounge. Richard enjoyed the work and liked the variety of working at different buildings and towns.

Richard was told that he had to go to college one day a week for the next three years to get qualifications in decorating and to learn all about painting. There was a lot to learn. Different surfaces needed different preparation and paints.

Richard learnt all about wallpapering different surfaces such as ceilings and walls, window reveals and chimney breasts. All aspects of spraying were covered, HVLP and airless.

The theory behind all aspects of decorating were covered in the classroom and he did many practical tasks in the workshop to perfect his skills.

Three years flew by.

In all this time learning about decorating, at work and at college, the process of how to price a decorating job or how to run a decorating business was not taught or shown. This was a shame because he quite fancied setting up his own business once he got into his twenties, and it would have been good to have had at least some guidance.

Of course, his boss had no reason to teach him how to be in business, Richard was a good employee and his boss wanted it to stay that way.

He could not complain though, the work was planned for him, the jobs were run by the foreman and he was paid every week without fail. Holiday pay was given when he was away on holiday, customers were dealt with and problems solved for him. He did not even have to decide what paints to use or which supplier to use.

Time moved on and Richard got lots of decorating experience. On his 22nd birthday he decided it was maybe time to have a go at working for himself. He had his own house now with a mortgage and bills to pay so it was a little risky, but he felt that he could earn more money and have more control over his time if he worked for himself.

He handed in his notice, registered with the inland revenue and bingo he was all set up ready to go. He decided to call himself "Richard Scarper Decorators", not very imaginative but it did the job, people knew who he was and what he did, so he felt that his name on the side of the van was a real asset.

He had obviously told friends and family his plan and he had lots of support and some offers of decorating work. Obviously, he was expected to do the work for a good price (cheap) because they were his friends and family. Some family members even expected the work to be done for free. He went along with this because it was work after all and he needed to fill his days.

Work was steady and enjoyable, it was nice to be able to start and finish when he wanted. He even had some days off during the week which was a novelty. One of the things that Richard didn't know how to do however, was price the work.

He had chatted to some of his decorator friends and they told him that he needed to guess how many days he thought the job would take and then multiply this by his "day rate" and finally add on the cost of the paint. That sounded fair enough, one or two of his mates were also working for themselves and this is how they did it so it must be right.

He had to decide on what his rate was going to be, everyone told him that £100 per day was a "fair" price and that is what he should charge. When he was working for the company, he was on £11.60 an hour. This gave him £464 a week before tax and deductions.

If he charged £100 a day then that would give him £500 a week. This was more money, so happy days. Some weeks he worked at the weekend, so he earned £700!

Wow that was a lot.

Richard was busy for a while and he even started to get some work from people that his friends knew. These people were a little bit harder to work for because he didn't know them. They also had been told by his friends how much he had charged for their decorating, so they expected the same price.

When Richard tried to charge more, they didn't like it and sometimes withdrew the work. If he was not careful, he was going to end up working on "mates' rates" for the rest of his career, this was no good.

Richard hit a quiet spell and he had no work. All his friends were done for now and friends of friends were not coming through with any work either. He decided to have a couple of weeks holiday to think it over. It was then he realised that there would be no holiday pay for this two-week period while he was not working.

Suddenly £500 a week did not seem enough.

While on holiday he realised his first mistake. He was a very good decorator and he did an excellent job and always left the job clean and tidy. He was also very honest and trustworthy. He was worth more than he had been charging.

His first customers had chosen him not because of these qualities but because he was cheap. They had lots of friends who also liked cheap work, so he got those as customers. He had started to build a business of cheapskate customers. This road was not a good one. Richard needed a re-think.

He decided to increase his day rate to £150 a day. This was quite a price hike, but he decided that it needed to happen so that he could at least afford to pay himself holiday pay. He also decided to advertise to attract a new type of customer.

He placed an advertisement in a local magazine that was delivered to local upmarket areas. He started to get some work from more wealthy customers. These people accepted his new day rate and before long he was making some good money. He felt that he had finally got his business on an even keel and was thinking about how he could increase his earnings. He had a chat with one of his decorator friends who persuaded Richard to take him on. Richard now had someone working for him.

Lessons learned

We learn a couple of things from this story.

Firstly, there are basic holes in the day rate approach, and you have to include more than just your pay. You must look at "overheads" too. We will look at this in more detail later in the book.

Also, it's important to understand the impact your pricing level has on the type of customer that you attract.

Look at the following brushes; -

Brush 1 – 50p
Brush 2 - £3.00
Brush 3 – £12.00

Which is the best brush?

Most people would say brush number 3. We use price as an indicator for quality. Customers will do that with you. If you are cheap, they will assume you are not that good. If you are expensive then they will assume that you are good. You need to be good though or else you will lose custom.

Chapter 3 – The most common pricing model – guessing

The previous chapter looked at what happens to most decorators at the start of their career. I know there are variations in reality, and it was only a story.

You serve your time at a company, as an apprentice you are not on great money, but you are learning the ropes. Then you come out of your time and carry on working for the same company. Then you start getting asked by friends if you will do some painting for them. You could do with a bit of extra money on top of what you earn at your company.

How do you price though? You are not sure, so you ask one of the guys that you work with. The words of wisdom roll off their tongue.

What you do is this.

You look at the job and guess how long it will take. Say 2 days. Then you guess how much paint you will need let's say 5 litres and then you multiply your days by the going rate which they tell you is £100 a day. Wow that's easy then 2 days times £100 plus let's say £50 for the paint. £250. It's a doddle this pricing lark. You thank your mate and give the price to your new customer.

The job does take 2 days and the customer pays you straight away. It's more enjoyable too because you have no-one checking up on you and you keep all the money for yourself. Maybe you could leave work and work for yourself all the time!

Hang on a minute..... you guess??

Really? Is that what your mate said. With a straight face. You bought it too without question. Granted at first you were not too sure but you asked a couple of other mates and they said the same so it must be right.

Anyway, we don't want to get too bogged down with pricing do we, it's not that important. You have all these amazing decorating skills that will make up for your lack of pricing ability.

Unfortunately, your decorating skills won't make up for your lack of pricing skills.

The ability to price is a skill in itself that needs to be learnt. You have learned how to paint, you have learned how to wallpaper and now you have to learn how to price. Just like any skill you will be a bit rubbish at it when you start but with practice you will get better and better at it. As you get better your business will blossom into something that is a joy to work in.

What's the problem with guessing you say, I have priced like that for years and I do well? There are a few problems with guessing, let's look at them.

1. The first problem is that it is not consistent.

This is one of the main problems. It's difficult to be consistent. If you priced someone's lounge and then they want a bedroom doing and it's a bit smaller how much less do you charge, or do you charge the same? Is the customer going to say that the bedroom price is more expensive or cheaper? If they say that are you sure that you guessed right?

Another problem is that it can depend on your mood. If you are feeling optimistic then you may think "oh yeah I will smash this in three days" or if you are feeling a bit miserable, maybe your last job took longer than you expected then you might say "I am going to price it at five days."

This is not a good way to operate. You are in a constant state of doubt in your mind, have I guessed right? What if I do it quicker? What if I do it slower?

I think sometimes you do it in the time that you guessed just so that you feel that you guessed right.

2. It's not scalable, it does not work on big jobs.

The next problem is this. You can fairly easily guess how long a lounge will take, you have done hundreds of them and they all usually take the same sort of time. If you guess at 4 days and it takes another half a day then it's no big deal. If you guess 5 litres and it takes 6 litres, then this is not really a problem either.

But then that big job comes along. It's a hotel and it has 300 rooms. All of a sudden you in a bigger league. If you guess half a day out with this one, then you are going to be 150 days out on the whole job. Plus, the numbers get scary.

Your single lounge was £450, 300 rooms is £135,000 phew that's over a hundred thousand quid, am I right with that? Sounds a lot! This is really when the guessing gets scary.

3. You can't price off a plan by guessing.

Another situation that might happen. You do the hotel job, lucky for you the price was right, and you have got a taste for the big bucks. Now the main contractor rings you up. They are building a new hotel in the next town, can you price it? I will send you the plans they say.

Plans?

What am I going to do with them? They arrive and they look complicated, guessing is not an option here, you can't even visualise what the job will look like.

4. If your company grows, will your employees work as fast as you. Will your guess be accurate for them?

We will look at this in the next chapter when Richard decides to take on a new decorator to help him out.

5. If you sell your company, will the new owner be as good at guessing as you are.

Ok we are 20 years down the line. Your company has a revenue of £800,000 a year. You have a good team, but you are still the only one you trust to price. Your company has been valued at a million quid. Wow happy days.

Only one problem though, no-one can buy your business and let you retire because without you nothing gets priced.

Potential value is a million quid, actual value is nothing.

Ok I know what you are thinking. You will never be in this position, so it does not matter. Maybe, but it's one of the reasons that guessing does not work well.

We will discuss this again later on in the book because it is actually more important than you may realise.

6. If you decide to employ an estimator then they will have to use some kind of system.

If you take someone on as an estimator then it is unlikely that they will guess the same as you, You would either have to spend a lot of time getting them to guess the same as you or you would have to develop a more robust method.

Just to finish off, I will confess that sometimes I guess how long a job will take and price accordingly. You do get very good at it and for some jobs, especially very small ones it can be the quickest and easiest way to price but we need another approach for larger jobs.

Chapter 4 - Taking someone on

The new guy was called John. John had been a decorator for 8 years and had always worked for a firm. John quite fancied working for his friend, he thought it would be a laugh and not like working for his old boss who was always trying to catch him out when he had too long for dinner or if he sneaked home early at 3:00pm in the afternoon.

From Richards point of view, John was going to be a great addition to the team. He knew John was a good decorator and he was his friend so he would not try and duck and dive. This would mean that Richard could take on more work and hopefully make more money.

John has said that he wanted £14 pounds an hour which was more than he had been earning at his old company, so Richard felt he had to offer this. Hopefully though now he was charging out £150 a day he should make some money on the work that John did. John was not "cards in" though he was sub-contracting. This still meant that Richard had to deduct tax under the CIS scheme.

The first job that John worked on was a large hallway stairs and landing and Richard was there too. They worked together on the job for a week and they made good progress. Having John there made the job progress much faster and by the end of the week they were done. Richard invoiced the customer and when he got paid, he worked out that he had indeed made more money than he would have done if he had done the job on his own.

That night Richard went to price another similar job. He decided that the job would take 2 weeks and he put in his quote.

10 days x £150 plus materials (£300) - £1800.

This was going to be a nice little earner. He told John where the job was and called the customer to let him know that John would be there at 8:00am in the morning. John arrived the next morning and started work. There was quite a lot to do, it was a large hallway, stairs and landing and it needed stripping and preparing before any decorating could be done.

The customer was supplying the wallpaper, and this had been quite expensive. John worked steadily though the week and was progressing well. By the end of the first week he had prepared everything and lined the walls. All that was left to do was the painting and the wallpaper.

On the second week Richard got a phone call from the customer on Monday morning. No John. He had not turned up on the job and he had not let anyone know. Richard called John and there was no reply. Richard decided to go to the job in Johns absence. John called later, he had been very ill in the night and had only just woke up, He felt better, and he would be on the job the next day.

No harm done and John came in for the rest of the week and worked steadily on the job. Thursday came and Richard was aware that the job had to be completed by Friday. He went down to the job to check everything was on track.

It wasn't on track.

"It's taken me longer to paint the hallway than you thought mate" said John.

"When do you think it will be ready for papering then?" Richard asked

"Probably Monday night."

"Next week!" exclaimed Richard.

"Yes, and I think it's going to take me 4 days to put the wallpaper on."

"That's an extra week over the price I have quoted, I am going to be losing money on this job" Richard said.

"Sorry mate but I think you have underestimated the time the job will take, I have been working my socks off for you."

Richard left the job and gave this some thought. He knew that he would have completed the work in 10 days, but I suppose John is not as experienced as he was, and he was probably working hard on the job. Richard did a new calculation.

John was costing him £110 x 15 days on this job.

So that was £1,650 plus the materials that had come to £340 in the end so that was £1,990.

That was a £190 loss on the job.

Not a good start.

John also reflected on the situation. He made himself a brew, well it was 3:00pm and he was not going to do much more today. He liked working on this job, it was close to home and quite cushy. The customer had been making him lunch too which was a bonus. No way he was going to take 2 weeks, he had decided to make it last 3 weeks. Richard was a mate, John was sure that there would be no repercussions.

The next job was a simple redecoration of a lounge. Richard felt that it could be done easily in 3 days. This time he told john that he would pay him 3 days for the job, £330. So, if it took longer then he would be out of pocket. John was not too happy about this but agreed.

Three days later John was finished, and Richard invoiced the customer. The new way of working had been a success and he patted himself on the back for being a great businessman.

That night Richard got a phone call. The customer was not really happy with the standard of work. The ceiling looked like it had only had one coat, in fact the customer knew that it had because he had been keeping an eye on John. John had left early for two of the days and gone home at 3:30pm.

Richard agreed to go around and put the work right and he spent another day on the job doing this. He also had a word with John who said he was sorry, but he was just not as fast as Richard and had felt that the only way to complete the job on time was to cut corners.

On the next job Richard felt that he could complete the work in 4 days, but he added on another 2 days because he knew that John would take a little longer.

6 x £150 = £900 plus materials of £180 = £1080

This was £300 more than Richard would have usually charged but he didn't want a repeat of the previous two jobs.

This seemed to work. The job was completed on time and the standard of work was good. Richard had gone around to check on the last day to make sure that the job was right. Richard had also visited the job at strategic times to try and catch John out when he did one of his early finishes.

Richard never got any more work from that particular customer though and this bothered him. He relied on repeat customers and he felt it was a good sign if customers came back for more work. He called the customer to see if he could find out what had gone on.

The customer was a bit embarrassed but went on to tell Richard that he felt that John had been more expensive than Richard would have been, and he also felt that John was not as good. He had always used Richard as a decorator because he liked him and felt he got good value.

If Richard was prepared to do the work in the future, then he would be happy to engage his services but if it was John that was coming to the job then he was not interested.

Lessons learned

This chapter looks at ways that a typical decorator looks to expand and make more money and although John seems like a nightmare in reality the situations that have cropped up are not uncommon in our trade. If you are going to look to employ someone you need to be aware of the pitfalls.

You can see also that the day rate multiplied by time method does not work as well once you introduce people other than yourself into the mix.

Finally, I want you to think about what happens when you decided to take someone on. It's not always as easy as it first seems. Most decorators do this to help them cope with an increased volume of work but also to help them make more money.

As we progress through the book you will see that there are better ways to expand and make more money.

Chapter 5 – Decorating is changing

I started my apprenticeship in 1982, back then I worked for a decorating company that employed about 30 decorators. We did a whole range of work and it was a good place to learn the trade because of this. We used brushes and rollers a lot and we also used oil based paints a lot. The main focus was on speed and not quality and we got the job done quickly so that of course the boss made some money.

Once I started teaching at college full time my mindset was locked into the 80's decorating way. I didn't really realise this until I went back on the tools recently. I was amazed how much had changed. Dustless sanding, airless sprayers and a whole new set of products that you could use to get a great finish.

Social media and YouTube have also changed the trade. We are able to look at what decorators are doing all over the world, we are able to share good ideas and practices and use this information to be better at what we do.

When teaching the apprentices at college I would show them some videos from a chap in the states called The Idaho Painter. Chris Berry does some amazing videos showing how he tackles various jobs that he does. One of the things that struck me about his videos, especially the ones where he painted the external of a property was the following; -

He sprayed most of the surfaces
He didn't get overspray everywhere
He masked very quickly
He only had a small team
He used water based paints
He had a system that he had thought through
He did the work incredibly quickly
He charged good money and made good money
He did a very nice job

At first, I found this quite hard to believe but the more I thought about it the more it made sense. I had taught spraying at college for years, the second year students learnt how to use High Volume Low Pressure systems or HVLP and the third year students learnt how to use airless.

Airless was incredibly fast and I had always vowed that if I ever went back on the tools then I would spray everything. Here was a guy who was already doing that, and he made it look so easy!

I started talking to other decorators that I knew, and they fell into a few camps. There were the old school guys who were doing the job in much the same way that we did back in the 80's. Brushing and rolling everything, using oil based paints and sanding by hand.

When I spoke to these guys I was usually met with a lot of resistance.

You can't do that Pete, you don't understand. It wouldn't work. I think we will stick to what we know works. I was a bit baffled by these replies but of course the proof is in the pudding and I started to take on contracts myself and put into practice some of the new techniques.

The younger generation of decorators coming through were more receptive to the new ways of working and were tech savvy too, so they were sharing ideas on Facebook and pushing forward with their businesses.

I bought myself some kit, a small airless sprayer, a masking machine etc. I also started to try out some of the water based paints that were now on the market, there are some good products out there and there is no substitute for experimenting with them yourself and seeing what is good and what isn't.

Everything seemed to come together at once. When I had used water based paints in the past, I was brushing them, and I found this quite hard and I struggled to get a decent finish compared to oil. However now that I was spraying, I found that the finish was incredible, and I could produce work at a much faster rate and the job was dry in no time too.

What has all this got to do with pricing?

Well as it happens, a lot.

You see the whole approach to the job changes. You now need to mask, and this takes time but once you start painting this is very quick.

For example, let's take a large window that needs 3 coats of paint.

The window will take about an hour per coat to paint by brush once it's prepped. So overall, it's a 3 hour job. Let's say you charge £15 an hour.

The cost of the window for your time is £45.

I come along and I mask the window in 20 minutes and then each coat takes 3 mins to spray. Now the job has taken half an hour.

Do I charge £7.50?

If the window needs another coat with the traditional approach then that's another hour, another £15 making it £60 in total. It's a big decision too, especially if you have given the customer a price. You would have to either swallow the extra time and not charge or approach the customer with this extra cost before going ahead.

Another coat when spraying would be another 3 minutes and a bit of product. No cost at all really to you. You can focus on doing the job right and giving it that extra coat if needed.

The traditional approach to painting means that you try and start painting as soon as you can in the day and work hard through the day to get as much painted as you can.

Time is money.

This does not happen when spraying, you have the opposite problem, all the surfaces are wet within an hour and you have nothing to do. I have started running 2 or 3 jobs together on the same day so that I have less down time and produce more work.

If this is not possible then I would take my time, make sure the preparation is spot on and maybe only start painting after lunch. This approach is quite hard for some decorators to get their head around.

You need to be more organised and think the job through before you start, much like Chris Berry with his systems. If you combine all these factors, quick drying paints, spraying instead of brushing and having a system them your productivity goes up by a factor of 4.

What does that actually mean though?

Well it means that you are producing 4 times more work in the week. So that could mean either that you work shorter days or fewer days or earn more money or a mixture of all three. I tend to go for the mix of all 3 and maybe work 2 or 3 days a week. Start at 9:00am finish at 3:00pm and earn double.

Now I know what you are all thinking at this point. You need to be careful that you don't upset the customer by earning too much and you need to be careful that prices don't start to fall as we get more productive. Builders for example are not stupid, if they see you earning too much then they will cut your rate.

This is one of the main reasons for writing this book, we all need to be clear on how we price and be able to manage the customer in a way that benefits us as well as the customer.

The good thing is that when you speak to Chris Berry about how he runs his business he does all the things that I am talking about and successfully manages the customer so that he can turn over thousands of dollars a week and turn out a lot of work. These guys have been doing it for years so it shows that it can be done without losing out.

You just need to get your act together.

One final thought before we move on. Customer's expectations are changing. They want a better job. I think we are lagging behind other countries with this but eventually expectations will rise. I think this can only be good for us as a trade and I will come back to this later.

In the next Chapter we are going to see what happens when Richard discovers spraying, how will it change his business. I think you may be surprised at what happens.

Chapter 6 – More productive methods

Richard decided to let John go, it was not really working, and it had put a strain on their friendship. John went back to his old firm. Richard continued working on his own.

He was on one of his many hallway, stairs and landing jobs when there was a knock at the door, he went down to answer the door only to be met by another decorator! Richard did not know the chap, but he introduced himself as Frank. Frank was here to spray the front door.

The front door was plastic and Richard did not even know that you could paint plastic let alone spray it. In fact, he now remembers the customer asking him about painting the front door and he had declined to do it.

Frank was a chatty decorator, he was relaxed and methodical. He prepared the door very well, he spent time to degrease it and abrade it. The masking took quite a while but not as long as Richard expected. The spraying process was amazing and once done the front door looked great. The whole job took Frank about 3 hours. An easy 3 hours in Richards opinion, Frank seemed to spend most of his time chatting.

It turned out that Frank was charging £250 for the front door, £30 of which was materials. This meant that Frank was making £440 a day! Wow that was amazing. After Frank left the customer came home and was over the moon with the front door.

No mention of how long it took. In fact, Richard heard the customer on the phone telling all his friends how "amazing" the finish on the front door was. He also heard the customer say that he thought the price of £250 was actually good value once he had seen the finished product.

Richard spent the rest of the day thinking about spraying and how fast it was and how great the front door looked once done. When he got home that night, he decided to Google spraying and see what he could learn.

Not many decorators did it in England he found but there were lots of videos of American and Australian decorators spraying. He also looked at the type of sprayers available and it seemed like there were loads and it was quite confusing.

The next day he bumped into John in the suppliers getting some emulsion. He told John about spraying and was surprised when he heard Johns reply.

“It’s only for big jobs mate, it’s very messy and you tend to get overspray everywhere, I had a mate who sprayed the outside of a house and all the cars in the street where covered in paint, cost him a fortune to put them right” John spouted.

The supplier agreed “We sell sprayers here, but they are expensive, you are looking at over a grand to get a half decent one, we don’t sell many.”

Richard was a bit discouraged. He remembered Frank had pulled up in a brand-new Mercedes van and during his many chats had not mentioned “overspray” or “large jobs” in fact he parked his van right outside the house while he sprayed the front door.

That night Richard was on YouTube to see what he could learn. There were some really good videos on spraying and one or two “how-to” videos too. This encouraged him. One of the videos showed a decorator using a small airless sprayer to spray a garage door. It looked like fun. That night Richard decided to buy himself a sprayer.

He went to his local supplier and bought a little sprayer for just under £1000. The supplier let him put the cost onto his account and pay it off over the next 3 months.

Happy days.

As luck would have it, he also had chance of painting a new house. This was all bare plaster and would be ideal to spray. He went to price the job and guessed that it would take him a week to complete.

He started on the first day and got his sprayer set up. It was very daunting, but he got it going. The sprayer did seem to produce a lot of over spray but as luck would have it, he bumped into Frank at the site office.

“Fancy seeing you here” Richard said.

He went on to tell him about his spraying journey. Frank was impressed at the moves that Richard had took saying that most decorators did not want anything to do with spraying. He came down to the house and gave him some hints and tips to get him going. Richard realised that the overspray had been the result of his spraying technique and now after Franks help, he was getting much less.

Frank also told him that he could spray the woodwork, this was much faster than painting by brush and it had the added advantage that the finish was amazing. Richard had a go at this too. He ended up brushing the final coat of gloss but at the end of the job that he managed to do in 4 days instead of 5 he was feeling pleased. For the first time he felt he had made some progress.

Richard reflected on the job and realised that although he had priced the job for five days he knew that now that he had sprayed it, he could easily do it in 3 days.

This was a problem for his day rate pricing system.

Maybe he could increase his day rate so that he got the same money as he did when he brushed and rolled. Then he thought that really, he should charge more because the builder was getting a superior finish. He had to admit that he didn't really know how to move forward.

He knew that the day rate system was no good when you started getting more productive.

The builder was having similar thoughts. That decorator had been very fast painting that house and he had made a brilliant job. However, there was no way he was paying him more than the standard day rate, he didn't care how good he was.

The next day the builder rang Richard with some more work. The conversation went something like this.

"Hi, is that Richard? I have 5 more houses for you to paint."

"Great" says Richard.

"How much?"

"Well I would charge the same as I charged for the one, I have just done for you, so that's £750 plus paint, £900" Richard replied.

"Mmm, how long do you think a house will take you to paint?"

Richard knew that he should not really answer this, but he was an honest and open guy so he said "I got quite good with the sprayer on the first one so I think I could do one in 3 days if I have 5 to go at."

"3 days……. £900……. I don't think so" the builder replied. "I will pay you "£600."

"Ok then" Richard replied rather easily.

Richard worked hard to get the work completed on time, harder in fact than before he got the sprayer. He was starting to regret the road he had started to go down and didn't accept any more work from the builder.

Back to the drawing board.

Richard decided he was going to stick with domestic work and advertise more and build a team. Not friends this time. He put an ad on Facebook for decorators and started to build his team.

Lessons learnt

This clearly shows what can happen when you try and get more productive with your decorating. Richard is lucky because he has Frank who helped him avoid all the common pitfalls to spraying however the builder has bullied him into lowering his prices and he has ended up in a worse place. This is one of the main reasons that I have written this book.

I mainly spray all my work myself and that is why I have included it into the book. Spraying is starting to become more popular amongst decorators too.

However even if you never touch a sprayer in your career you will still gain a lot from some of the ideas that I will cover in the later chapters.

Just bear with me.

Chapter 7 – Supply and demand – A simple lesson in economics

The previous chapter got to the heart of the problem.

We admired Richard for trying to get more productive and therefore make more money, but we were disappointed that the builder got the better of him and cut his price.

In the builders' defence he was just doing his job of trying to get the best deal from the decorator and keeping costs down so that he made more profit. The builder was in a strong position because he had a lot of work to offer Richard.

Richard on the other hand should have been a bit wiser with his approach to the builder and pointed out the **value** of what he was giving the builder. He was giving the builder a *better finish* at no extra cost and he was also helping the builder finish the job on or *ahead of schedule.*

Before we get into the actual pricing of a decorating job I want to talk about supply and demand and cover a bit of basic economics. You may be tempted to skip this chapter and cut to the chase, but I am asking you not to.

This is very important stuff that you need to understand if you are going to be a black belt in pricing.

How do Ford decide on a price for their Mondeo? Do they guess how long it will take to make and multiply that by £100 a day and then add materials?

No of course not.

Do they look at how long it takes to make a Mondeo and use that as a basis for their price. Well yes, they do this, they know how much the car **costs** them to make. I don't know how much it costs to make a Mondeo but let's guess at £6,420.

If Ford sell the Mondeo for £20,000 then they make a fair bit. Obviously in reality there are dealers etc who get a cut.

Even this is not the way the price is set.

If everyone stopped buying cars this year and the demand for cars went through the floor, then the market price for the Mondeo would drop. Ford of course would make less profit. If Ford dropped the actual price then people might start buying them again.

The price of anything is set by how many people want to buy the product compared to how many people are selling the product.

Houses are a good example because their price changes all the time, so let's look at those.

You own your own house and you decide that you want to sell it. An estate agent comes around to have a look and also, they look at similar properties in your area and they give you a price.

£180,000 they say.

Not bad you think I could cope with that. Just to be sure you get another estate agent in. They say £200,000. Even better! We are going to go with that estate agent!

Hang on a minute what's going on here?

Well what's going on is that until you put your house on the market, no-one really knows what the house will bring so the estate agent makes their best guess based on prices that other similar houses have sold for in the last few months.

The estate agent with the higher price might just be wanting to get your business. You might sell at that price, but it may take 9 months. At the £180,000 price it may have sold in a couple of months.

Your house might be a one off, a special property that you built yourself. There is nothing like it in the area. **This would be harder to put a price on.**

Notice that nowhere in this conversation do we discuss how long the house took to make.

One last example is the auction. The marketplace in action. You decide to auction your house off because you want a quick sale.

The room is full. The bidding starts.

Who will give me £150,000?

A hand twitches.

£160K?

Someone else jumps in.

£170K?

A third person indicates their interest.

This goes on and the hammer goes down - £216,000.

Happy days.

The funny thing is, the day after it could have sold for £190K or £210K depending on who was in the room and how much they wanted your house. The higher the price goes the less people are interested and the lower the price is the more people are interested.

If the price had started at £50K then probably everyone in the room would have put their hands up.

If the price started at £300,000 then no-one would have bid.

Supply and demand in action.

Lower the price and get more demand, raise the price and get less demand.

How does this work in the world of decorating?

It works like this. Let us assume we live on a little island and there is only 1 decorator, let's call him Dave. Everybody wants this one decorator. He cannot work for everyone because there are only so many hours in the day, so he sets his day rate at what he thinks is fair.

The rate is £150 a day.

At this price some of the people on the island decide to do their own decorating because they cannot afford the price. The problem is that there are too many customers at this rate. The rate goes up.

£180 a day.

A big jump, this puts quite a few off but not everybody. Dave is still pretty busy, so the price goes up again.

£300 a day.

Bang. No work. The price has gone too high and no-one will pay it. Finally, our island decorator drops back to £160 and finds nice steady flow of work.

This is the demand side of things.

The one decorator is the only supply and the demand changes depending on the price. But in the real world it is more complicated than this. When the day rate goes up to £160, someone else on the island who is quite a dab hand at decorating decides to enter the market. Let's call this decorator Debbie. Debbie does a course at college (ok it's a big island) and starts advertising.

Now at this stage Debbie is only really learning and is not as good or fast as Dave so there is no real threat. To combat this Debbie goes in a bit cheaper. At £80 a day Debbie starts to get some work. Some of the customers are Dave's. Over a period of time Dave starts to get quieter and has to drop his price. Not down to £80 but down to £140. Some of Dave's customers like Dave and know that he is reliable so will not switch easily.

This is the supply side of things.

More people decorating will push down prices that decorators get.

Demand for your services will change if the price that you charge goes up or down and prices will change depending on how many people are actually decorating.

The speed at which demand changes the price is called *price elasticity.*

There is ***elasticity of demand*** and ***elasticity of supply***.

1. Price elasticity of demand

This is how quickly demand for a product changes when the price changes. For example, petrol has an inelastic demand. If the price goes up, then people still buy it because they need it and there are not many alternatives.

In the short run people may car share or cycle but in the long run they may convert their cars to liquid gas. The government tends to pick goods like this to tax because they know that slapping 80% tax on petrol is not going to stop people buying it.

Crisps on the other hand have a very elastic demand. If Walkers double the price of their crisps, then people would stop buying them immediately. They would just buy Seabrook instead or maybe an apple. There are lots of alternatives.

Decorating has quite an elastic demand. If prices go up, then people won't buy because they can live without their decorating being done for a while.

2. Price elasticity of supply

This is how easy it is for people to enter the market when the price for that skill goes up because there is more demand for it.

Plastic surgeons are a good example.

When the first surgeons started doing plastic surgery, they could earn millions of dollars every year. This is because there was a big demand for their services and there were only a handful of people doing the job.

Young people leaving school would see that plastic surgeons earned $2 million a year and think, I will have some of that. They go to medical school and train. This takes years but eventually they start to practice. Quite a few would do this so that there is an increased supply of surgeons.

Currently in the USA plastic surgeons earn $273,000 per year, a lot of money but less than the millions that were previously earned.

The elasticity for supply for plastic surgeons is inelastic because it's hard to get into the job. You have to be clever and good with your hands and the training takes years.

Another example might be someone who sets up washing cars. They buy a bucket and sponge and they are in business. If the price for car washing goes up to £50 a car then people would set up doing this very quickly and the price would drop back down. The price elasticity of supply in this example is very elastic.

A lot of businesses have what are called barriers to entry. A barrier to entry might be the fact that you need a medical degree before anyone will pay you to be a doctor. Another one may be that you want to set up as a printing company, but you need to buy printers and rent a property.

Barriers to entry create a more inelastic supply because it's hard to enter the industry. Some industries will create barriers to entry to protect their own income. Some professions limit the number of people that can enter the profession. This is getting more difficult these days as markets become more global and you can do a lot of things using the internet.

This is one of the problems that we have as decorators, people perceive that it is easy to enter the market. They think they can buy a brush and roller and a pair of steps and they are in business.

There are some barriers to entry. To be good at what you do requires skill that takes time to get really good. You need to gain a qualification so that you can show this to employers and customers. You need to buy some equipment and a van.

In reality though you can just set up, call yourself a decorator and go for it. If you're a bit cheaper people will not bother too much as long as your work is average.

This means that in theory the price for our skill can be quite volatile and change quite quickly depending on how much work is out there.

However, there is a way that you can bypass this supply and demand process.

Let's have a quick look at how.

George Clooney earned $239 million in 2018 before tax, this is not a guess, I Googled it. Take a look on Forbes 2018 list of top male earners.

Now why is that?

There are thousands of actors in the states if not hundreds of thousands. They are not rare. You don't need any qualifications to do the job and you don't have to be particularly clever. I am not having a go at actors it's just a fact. Most actors probably earn minimum wage.

The difference with George Clooney is that although he is an actor and there are thousands of them there is only one George Clooney. He is unique. *He has created his own market with only him in it.* This means that no-one can enter the market to compete with him. He can ask for increasing amounts of money as more and more companies want him in their films.

Anyone can do this.

If you advertise yourself as a painter and decorator, then you are in danger of being lost in a sea of people doing the same thing. However, if you specialise in a certain type of work for a certain type of customer and then get well known in this field then customers would start just wanting you and you would have a more stable income.

This is not something that you can do overnight but will take a while to build up, but once you have you will be in a strong position when pricing.

This is not my own idea and it's something that you could write a whole book on. In fact, someone has.

Check out Daniel Priestley's book called Oversubscribed. It's well worth a read.

Our hero Richard is going to go down this road so I hope you will understand better how it can work for a decorator.

Let's go and see what he is up to now.

Chapter 8 – Learning from other trades

When Richard looked back on this period of his career, he only remembers hassle and more hassle and not much more money.

He had done the maths.

If he charged £150 a day and paid £110 a day then he would make £40 per day per decorator. If he had 10 decorators, he would make £400 a day, this was £2000 a week.

This was the big time as far as Richard was concerned.

So, this was his plan. All he needed was enough work for 10 decorators and also of course 10 decorators to do the work.

Over the next 4 years Richard expanded his business and took on more staff. He had many problems along the way. Getting staff was the first one.

"I have seen your advert for staff on Facebook and I am interested in a job" said Joe.

Joe had been decorating for 4 years, he had never been to college and had only really done site work, but he thought that "Richard Scarper Decorators" sounded a good place to work.

"Great" said Richard "Meet me at the job on Monday morning and we will see what you can do."

That was the last Richard heard of Joe, he didn't turn up and didn't answer his phone again. Strange.

"I have seen your advert for staff on Facebook and I am interested in a job" said Andy.

"Great" said Richard "Meet me at the office on Monday."

Andy did actually turn up at the office on Monday. He was a little late though. He said that he was an experienced decorator and could even wallpaper. He would not however work for £110 a day, he was worth so much more. £230 a day even.

Andy did not get the job.

"I have seen your advert for staff on Facebook and I am interested in a job" said Ben.

Ben turned out to be alright, he was a decorator who was happy with the rate of pay. Ben however seemed to like to have Monday off and sometimes

Friday. He had various reasons for this. His nan died (4 times) he had a flat tyre on his car (17 times) he had overslept, and it wasn't worth coming in (22 times) and he had gone on holiday and the flights were delayed. This was used once so it was probably true. Ben did not work out and had to go.

Richard did build up a team of 10 decorators but boy it was hard work. It was like herding cats. It was hard to keep the flow of work going too. Richard spent all his time pricing new jobs, invoicing jobs, firefighting problems on jobs, sacking people, hiring people, organising people. One decorator had fallen off a ladder and was in the process of suing him.

At the end of the year after all the costs had been taken out, he found that he had made hardly any profit after he had paid himself a wage. Richard was not decorating anymore, and this new job role was better paid but not enjoyable at all. Richard decided that he was going to pack up and just get a job like everyone else and at least then he could sleep at night.

His mate Charlie was a carpet fitter and was really busy working for high end clients. He needed another fitter and was prepared to train Richard and pay him £600 a week. Just 9 to 5 and no hassle. A regular pay check.

Happy days.

Richard looked back on his carpet fitter days with fondness. Charlie was a great guy, very organised, very well paid and he clearly had a good business. Richard knew almost straight away that what he really wanted was his own successful decorating business but that he had gone about it all the wrong way. It was good to speak to Charlie about business because he had a completely different perspective on everything.

One day while they were having their dinner the conversation got around to pricing.

“How much do you charge to fit a lounge carpet Charlie?” Richard asked.

“I charge £10 per metre” he said, “So this lounge today is 8m by 10m, so its $80m^2$ times £10.”

“800 quid” Richard gasped “How long will it take us?”

Charlie frowned, unsure what to say. “Erm why do you ask?”

“Well how did you work out your price? What is your day rate?” Richard pressed.

“Day rate? What are you talking about? I price by the area of the floor, so much per square metre, that way if I am more productive, I make more money” Charlie said.

“But you must know how long it’s going to take” Richard said.

“Look mate, everyone charges by the metre, if it’s a high end product and a high end customer then that rate is even more, nobody ever asks me how long it will take and even if they did, I would not tell them because it does not matter. I charge by the metre” Charlie explained “I have 4 fitters including you and I make around £100,000 a year.

My fitters make really good money too and I do not have any problem getting good fitters with the money that I pay them. My customers are happy, and they are happy to pay me because they know that I do a first class job.”

Charlie went onto to explain how he had built his business to the place it was today. He had decided at the outset that he was only going to work for people who had houses worth more than a million pounds.

He then spoke to these potential customers to find out what they were looking for in a carpet fitter. He learned that they valued a first-class job, a competitive but fair price and someone who was well presented, reliable and polite.

Charlie then set out to build this image for his business. He had new vans that were clean and well presented. He had a company uniform that he and all his fitters wore and had a business system that made sure that he was prompt and efficient.

It had taken a while to develop this niche of the market, he had turned down many well-paying jobs that did not meet his million pounds property criteria however in time he got the reputation with his wealthy customers as the "go to" guy for carpets and flooring. They told their friends how good he was and eventually he had a good stream of work meeting his criteria.

All his marketing was aimed at his selected group too.

Charlie explained that there was a difference between what it cost him to fit a carpet and what the value was to the customer. Even though you could work out the cost for yourself so that you knew your minimum price it was not to be used as a basis for your actual price.

Your actual price is based on the value of what you provide to the customer.

Richard remembered when Frank had sprayed the front door for the customer. Although Franks costs where maybe £100 for his time and materials the value to the customer was much higher. It would have cost the customer £1000 to get a new front door, so £250 was actually fairly cheap. Richard had never considered what value he gave to his customers, but he realised that he needed to think about this.

Charlie went on to say that it was important to **know your value** and be able to hold your own in the pricing negotiation.

"Negotiation?"

Charlie went onto explain that you needed to educate the customer about the value you gave and make clear that your price was set. If they wanted a cheaper price then there had to be some concessions for example a cheaper carpet or cheaper underlay, he would never just drop his price for no reason because this would signal that Charlie did not really value his service as much as he said. It would also mean that when Charlie was recommended to the client's friends, *they would also pass on the fact that Charlie would drop his price after a bit of pressure.*

Richard remembered the phone conversation with the builder and realised that he been a bit weak and should not have easily dropped the price without any concessions.

Maybe he should have held his ground and reminded the builder of the value he was giving. He should have reminded him that he had a programme to follow and completing the houses quicker meant that he would meet that programme.

He should have reminded him that the finish he had produced was far superior to the finish he had had in the past and that Richard had not charged any more for this superior finish, in short, he should have negotiated the price better. If he had he would have probably ended up doing the whole site at the original £900 per property.

He enjoyed chatting with Charlie about business and it was Charlie himself that said "You need to go back and have another go Richard, I think you are wasted as a carpet fitter, I know how good you are at decorating, but I want you to meet a good friend of mine who has been a great mentor for me and my business, his name is Martin. If you are free tonight we could go out for a drink and a chat, I regularly meet up with Martin for a drink and to talk business"

Charlie went on to explain that Martin was a one of his millionaire house owners and that he had fitted carpets in almost every room of his lovely house.

Martin was only 32 and had been a self-made millionaire since the age of 20. Richard found this fact alone amazing and he was keen to meet Martin and hear what he had to say.

Lessons learnt

Building a big decorating firm is not always a good thing if it's not done in the right way, and can just end up multiplying your problems.

It's important to price on the value that you give to the customer and use a per square metre rate for what you do.

It's important to understand the value that you give and explain it clearly to the customer so that they do not undermine your price. Be prepared to walk away and stick to your guns.

Notice also that the carpet fitter researched his chosen market to see what they were looking for and then gave it to them. Don't just assume that you know. This is how to build value.

I know so many decorators that turn up in a scruffy van wearing scruffy overalls and use curtains for dust sheets and then wonder why they cannot attract high end customers. Please don't be that person.

Chapter 9 – Positioning and negotiation

We covered a lot of ground in the previous chapter. Richard had met someone who approached their business in a completely different way. First of all, they priced by square metre instead of by day rate.

A lot of other trades do this, not just carpet fitters. Electricians will price per socket fitted. If someone asked them how long, it took to fit a socket they would not really know because in fairness its more complicated than that.

Plasterers will charge and be paid by the square metre. Joiners may charge so much to hang a door. I hope by now that you are starting to understand that you need a better system for pricing. I will get to some suggested pricing systems later on in the book for now we have one or two other issues to address.

It's no good getting your pricing right if your business is not well positioned and you cannot negotiate with the customer.

What do I mean by positioning?

The carpet fitter in the story had decided that he wanted to work for a small set of customers who owned very expensive houses. The decorating industry has many different types of work, let's explore what is out there before we discuss positioning.

First of all, there is the split between domestic and commercial. Some decorators do both.

Domestic decorating is where you deal directly with the customer and paint their house, either inside or outside. In this market there are many different types of people. There are the very wealthy who just don't have time or inclination to do the decorating work themselves. There are the middle class, doctors for example who can afford to pay someone else to do their decorating for them.

There are older customers who are not physically fit enough to climb a ladder and paint the outside of their house or do their hallway stairs and landing. I could go on.

The advantage of domestic work is that it tends to be more rewarding, the work is done to a higher standard and you get some job satisfaction at the end when you see the customer is pleased with the work that you have done. Domestic customers can be more loyal and if they like you and your work then they will keep coming back when they need some work doing. They will also recommend you to their friends.

The disadvantage of this type of work is that it tends to be lower volume and lower value work. So, for example it is very unlikely that you would get a £100,000 decorating job for a domestic customer. Possible but unlikely. This brings us to the next type of work which is commercial.

Examples of commercial work are new build houses, apartment buildings, shops and factories. There is also local authority work which is schools, hospitals and libraries. This type of work tends to be bigger and have a higher price. Commercial jobs can quickly get to £50,000 or £100,000 the sky is the limit really depending on the size of the work.

The advantage of this type of work is that it is easier to get a continuity of work, you may have a big job and you know that it will take 9 months to complete for 4 decorators. This makes employing people more stable. Another advantage is that it can be very lucrative. A £100,000 contract with a 20% profit margin is going to give you £20,000 profit if it all goes well.

The disadvantage is that the market is more competitive, and the standard of work tends to be lower. Many decorators do not like to work on big commercial work and prefer the domestic jobs.

If you are a decorator and have been working in the industry for many years you will be aware of the different types of work out there. When decorators set up on their own and get themselves a van they will typically put "Domestic and Commercial" on the side of their van. Sometimes they will have "no job too large or too small."

This is fine if that is what you want to do but you may find that the different types of work need a different skill set and a different pricing strategy so you may find you are just average at everything.

What you should do is sit down one sunny Sunday afternoon and have a think about the specific type of work that you want to do. What type of customer do you want to work for? What type of work are you good at?

Remember that there are advantages and disadvantages with each type of work so there is no ideal decorating work. *The main thing is that you can do the type of work really well.* The main aim is to make yourself different from the average decorator out there so that when customers are recommending you, they focus on your main strengths.

Here are a couple of examples.

You may be very good at hanging wallpaper and you have found that many decorators shy away from this type of work. **This could be your niche.** You may also decide that you will specialise in wallpapering for hotels.

You would research your market, find out which hotels have wallpaper in their rooms and in the common areas. Then you could either approach the hotel directly or approach the main contractor that deals with the hotels and offer your specialist service. You could even call yourself "Wallcovering specialists" to set yourself out from the crowd.

When I went part time from teaching at college and went back on the tools, I decided that I would specialise in spraying. I actually have quite a few skills at my disposal, I can signwrite, I can grain and marble, I can hang wallpaper, I can gild. However, I wanted to differentiate myself in the market and just focus on one area.

So, I advertised myself as a specialist in airless spraying, coined the term "Fast and flawless" and put myself out there. I got really good at getting a great finish and people started to talk. They didn't come to me because of price they came to me because of the sector of the market I was in.

It was still important to be fair and give a really good service but all of a sudden, I had more work than I could handle and that meant that I could pick and choose the work.

The next thing I did was focus on the type of customer that I wanted, people that were happy to pay me a deposit and then pay me on completion.

This is what it means to position yourself. If someone asks me to wood grain a garage door then I pass this onto another decorator friend who specialises in this kind of work.

People don't really see me as a decorator they see me as a sprayer. They have no real idea what a sprayer charges, so this makes it easier for me when I price jobs.

There are thousands of niches that you could carve out for yourself. Just choose something that you are good at or something that you are prepared to work at to get really good so that you become the one that people go to for that kind of work.

I could write a whole book on positioning but there are already books out there on this topic. Have a look on Amazon and see what there is. Not many decorators will do this so it will set you apart from the crowd.

The other thing I want to discuss with you is negotiation. We all know what this is, but have we given it any thought? When it comes to giving a price to a customer there will be a certain amount of negotiation. Even if it does not feel like it.

Remember that the customer will judge your price by the value that they feel that you are giving to them. If they feel that the value is no different than every other decorator, then they will expect your price to be the same as the 2 other prices they have been given.

However, if you know what value you give to the customer then you need to educate them about this. This is part of the negotiation. You need to drop into conversation that you are unique and the finish the you give is outstanding. Any strengths that you have need to be subtly brought to light.

Honest, trustworthy, on time, clean, tidy and polite. All these things are of value to the customer. You know the type of customer that you work for, so you also know what they are looking for. Make sure that you remind them.

This is all done when you are looking at the work and giving a quote.

Then comes the closing of the sale or landing the job. Remember when Richard was on the phone to the builder and the builder knocked him down to £600. The builder was negotiating.

In this situation you need to stand firm and be prepared to walk away. Remind them of the value that you give. If they insist on dropping the price, then you need to have a reason to do this. One coat instead of two. A different product. Personally, I would not even go down this road but that is up to you.

If you do drop your price for no reason other than you want the job, then you can bet that the customer will pass this fact onto anyone that they recommend you to and you will be constantly asked to drop your price.

This is one of the reasons why it is good to have a pricing system. You can then say that this is what the price came to when you calculated it.

Obviously, you need to keep reviewing your prices. If you don't get any of the jobs that you price, then maybe you need to re-examine your system. If you get every job that you price for then it is very likely that you are too cheap, and you need to look at this as well.

In the next chapter Richard meets someone who teaches him how to take his business to the next level. I have included this just to make you think about what is possible and where you could be if you get the formula right.

Chapter 10 – Getting some advice

Richard walked into the local pub where they were going to meet. It was a lovely place, slightly out of town. It had a great atmosphere. Sat at the bar was Richard and another chap who he presumed was Martin.

Martin looked young and bright eyed. He was in the middle of an enthusiastic conversation with Charlie.

"This is him now" Charlie said, "I was telling Martin all about you and your decorating business."

"Hi guys" Richard did a pathetic wave.

"So what business are you in Martin?" Richard asked. He looked like an Accountant or maybe a software guru, he was sure that whatever Martin did it was glamorous and very highly paid.

"I sell balloons" Martin replied with a sparkle in his eye "Not what you were expecting eh?"

Richard was lost for words, balloons, really, what could this guy teach me? A balloon seller?

He went on to explain that he started when he was 18 years old selling balloons at an amusement park.

He bought the balloons for 5 pence and sold them for £2. On a good day he would sell 500 balloons. The next step was to employ other balloon sellers. He paid them £200 per day but of course he himself made much more than this. Before long he was making a cool £20k per week and by the end of that year he was a millionaire. The company has expanded nationally and last year he sold it for £45 million. He was between businesses at the moment and looking to invest in a new venture. The decorating business sounded promising.

"Maybe I could invest in your decorating business Richard?"

"Well, maybe you could but I only really make a wage from the decorating I am not sure it would be worth your time and money" Richard replied.

"That is your first mistake Richard, you are not in business to make a wage, you are in business to make a profit and that is a different thing entirely, before I came here tonight, I did some research on the decorating business and in my opinion, it is ripe for change. Any sector that is ripe for change has profit in it."

Martin went on to explain that Richard needed to structure the business so that it made a profit over and above the wage he was taking out of the business.

"How much did you pay yourself when you were decorating?" Martin asked.

"About £28k a year."

"That is not enough, you are no longer a decorator, you are the managing director of a company and you need to be looking at earning at least £75k a year, maybe even £100k" Martin replied.

Richard was a bit taken aback by this. None of his decorator friends earned £100k a year not even close. In fact, he didn't know anybody who earned that except maybe Charlie. It did get him thinking though, he had been so focussed on decorating that he had lost sight of the fact that he was in business. If he was going to come away with anything tonight it was that you needed to make a profit and not just a wage.

"How do you price your work Richard?" asked Martin.

"Well I guess how long the work will take me, then I multiply that by my daily rate. Then I add materials and now after this conversation I am going to add 20% for profit" Richard replied.

"Mmm, I don't think that is a good pricing strategy, it's maybe ok to find your breakeven price but no good to find the price that you should be charging" Martin replied.

"I am not sure that selling balloons is the same as a decorating business though, you are buying and selling, there is no skill involved" Richard countered.

"Point taken, however I have a good friend who is also a millionaire and he provides a key cutting service. The key blanks are very cheap, about 1 penny each if you buy them in bulk. The labour cost to cut a key, (it takes seconds to do) is probably about 15 pence. So, let's say he uses your pricing model. He would charge 15 pence labour plus 1 penny for materials and maybe 4 pence profit. That's 20 pence for 1 key cut.

Does he charge this?

No, he charges £6.99 which is a whopping 40 times more. He cuts about 150 keys a day and this makes him about £1000." Martin replied "you see the value of the key to the customer is much more that 20 pence, it's more like £6.99 so that is what they are prepared to pay.

It's not the cost of what you do that sets the price it's the value to the customer"

Richard thought about Frank and his front doors. Frank had been very laid back and chatty so he was sure that if he put his mind to it then he could do 4 front doors a day. This would be £1000 less the cost of the paint. He remembered Frank saying that he did not use much paint. The customer did value what Frank had done and was over the moon with the amazing finish on the front door. In the customers eyes he had saved £750 on not buying a new front door.

"Your value to the customer is dependent on a number of factors, how good you are, how unique you are and what others charge for the same service" Martin explained "You can differentiate yourself in a number of ways, but one way is to become a specialist. For example, Charlie specialises in high end carpets for million-pound house owners. This makes him different from other fitters and means that he can command a higher price."

Richard started thinking about this, it shed a whole new light on what he had been trying to do. He needed to think about specialising and then calculating the value of that niche to the customer.

He had an idea that he could specialise in spraying shop front shutters. These could be sprayed, and he could source a specialist paint.

He felt that the value of this to the shop keeper was quite high and that he could charge between £1000 and £1500 for a shop front. The more he thought about this the more specialist niches came to mind, in the end he had too many to choose from.

"Why did you decide to set up your own decorating business Richard? What was your end game?" Martin asked.

Richard was a bit puzzled by this question.

End game?

What was that? He had simply gone into business to pay himself a wage and maybe have a little more freedom. This is what he told Martin.

"That's a good answer, of course we all want more control of our work lives, I always find it amazing that people go on about how important it is to own their own house but never mention how important it is to own your own method of making a living.

You can put your heart and soul into a job for 40 years making your boss very wealthy and end up owning nothing of what you have produced.

The value a business can be worth after even 5 years makes house ownership seem like small change." Martin said.

He went on to explain that you needed an end game too. What did you want to do with the business you have built when you came to retire? Do you want to pass it onto your children, or maybe do you want to sell it?

You don't want to just shut up shop and stop, that would be such a waste of all the work you have done over the years to build a business. Whatever you decide to do the result is the same, you need to make the business so that it can run and make profit without you and then finally be sold.

Richard had never thought about it this way, he had always been the exact opposite, his business needed him, he was the brains and the brawn, he priced, built relationships with customers, did the work. He was skilled, he was special.

None of this was any good if you are going to build a business to sell. You needed a business that you could train someone to do, you needed a business that was scalable, and you needed a business that had processes and procedures that meant that it would run on its own Martin explained.

"When I sold my business for £45 million, I was making 10 million pounds profit a year, the offer for my business was only 4.5 times the annual profit and that was because it was a relatively simple business.

However, the reason I could sell it was that I had people and procedures in place which meant that I was not needed at all for that profit to come rolling in."

"What do you mean by scalable?" Richard asked.

"It means that you can expand your business by duplicating what you do in other towns and cities. For example, with the key cutting service, once you had worked out the process and systems that say generated five thousand pounds a week in one place you could replicate it in 10 other places and generate 50 thousand pounds a week. This is £2.6 million a year. The main thing is that you develop a system and this system is perfected so that you can duplicate it and finally sell it." Martin replied.

The three guys spent the rest of the evening talking about business and at the end of the night they all went their separate ways. Martin had said that he would help Richard have another go at the decorating for a percentage of the business and Richard said that he would think about it.

Lessons learnt

You need to make a profit and not just a wage.

You need to know the cost of what you produce but also the value to the customer so you can come up with a price.

You need to be able to scale your business if you want to make larger profits. This may only be in the future but you need to start laying the groundwork now.

You need to have an "end game", what do you want to do with your business when you no longer need it or you want to retire.

Chapter 11 – The end game

The previous chapter had Richard talking to someone who was already successful in business. We all take advice from people all through our lives, but do we really get good advice from people who are well placed to give it?

When our mates (who are in the same boat as we are) give us advice about our business do they really know what they are talking about?

On one of the spraying courses I was teaching at the weekend I had a young decorator who was bright eyed and keen to learn as much as he could. At the end of the course when everyone had gone, he hung back to chat about what was the best sprayer he could buy.

I asked him what his budget was.

“Unlimited” he said.

This got my interest, he only looked young and he had an unlimited budget, how could that be?

After chatting I learned that he had built a business with the intention of selling it within 5 years. He had done that, and he was now looking to set up something new. He recommended a book to learn more about selling a business. This was called "Built to sell" by John Warrillow.

Check it out it's very interesting.

This made me realise that not only should you aim to build a business that makes you profit, you should also consider what you are going to do with the business once you are ready to either retire or move onto pastures new.

This book is not about how to build a business to sell but if you are going to go down this road then you need to put a few things in place.

The business needs to be able to run without you. This means marketing, pricing, day to day running must all be able to be done by someone else. So, at the back of your mind you need to remember that if there is only you that can do something then longer term you need to teach someone else.

This includes how you price.

Whatever system that you develop, and we will look at a few later in the book, it needs to be able to be followed by a successor.

So many decorating businesses rely very heavily on the founder of the business and I can cite a few examples where the founder has wanted to retire and could not do because the business would fail without them.

This may not be what you want now but if you are successful over the coming years you may change your view and when you do maybe revisit this chapter and give it some thought.

In the next chapter Richard is going to follow the advice he has been given and build a successful decorating business.

This business is called Flawless Finishing Ltd.

Chapter 12 – Flawless Finishing Ltd

Richard had decided to start again. Martin had invested £100K so that Richard had time to sort things out without having the pressure to pay the bills, it also gave him some working capital to buy some equipment.

First of all, he needed a name for the company. He had decided to specialise in spraying garage doors. He wanted a name that did not include the words "painting" or "decorating" or his own name.

He did not want people to categorise him as a decorator because this would mean that they would also assume that he charged decorator prices, he wanted to create his own niche so that he could control the value that the customer perceived. He did not want to use his own name because this would make the business less valuable when he eventually came to sell it.

The name he came up with was "Flawless Finishing Ltd", the good thing about this name was that it did not limit him to spraying garage doors. He could expand into doing shop shutters too for example.

Next, he decided to find out how much the customer valued having their garage doors refinished. He found that most people where happy to pay £250 for a single garage door and £325 for a double. Then he experimented with products and systems and found that with a proper sander he could prepare the door in half an hour and clean and mask it in another half an hour. Spraying took minutes and the whole process could be done in 2 hours. This meant that on a good day you could do 4 garage doors and if they were double then this did not really take any longer. Four doors would bring in £1,000 per day.

He also decided on a shift system so that there were finishers working 7 days a week. This would mean that people who wanted the work doing while they were home could be accommodated. This now meant that Richard was taking £7000 per week.

Richard also took the decision to call his operatives "finishers" and not decorators. He contacted a private training company and asked them to devise a training programme for his "finishers" and this would mean that they had professional training and also gained a recognised qualification.

The private training company was assured by City and Guilds and they were able to put the City and

Guilds badge on the certificates that his operatives achieved. This meant that he could advertise that all his finishers were fully qualified. He also found that the people he employed valued the fact that he was investing in training them.

Another staffing decision he made was not to recruit anyone under the age of 20 and anyone who was already a qualified painter. This seemed like an odd decision to make as Richard knew a number of good decorators that would be willing to work for him.

He had decided that he wanted to train his operatives up from scratch and wanted them to follow the system to the letter and not cut corners. Painters he felt would be tempted to cut corners. For example, they may not spend the time needed on preparation and try and get 5 or 6 doors done in the day.

Richard decided that he was going to pay his finishers £40,000 a year. They would not be on day work or a price, but they would get a bonus if they hit the target of 20 doors a week. This was a fairly easy achievable target, and this would get them a £200 bonus. The customers would have to fill in a feedback sheet and the bonus was dependent on 5-star feedback on the jobs that week.

The right type of people to be finishers turned out to be people who had worked in the retail industry, they were in their 20's and wanted an opportunity to get a skill and earn good money. The female finishers turned out to be the best of all.

Everyone wore the company uniform, and everyone was trained to handle the customer if they were on the job while the work was being done. "Giving the 5-star experience" booklet was essential reading for all new finishers.

Richard also wanted to build a good customer base. He was aware that some customers were better to deal with than others and he developed a system so that customers paid a 50% deposit before the work would be booked in and then they paid the balance on the day of completion to the finisher. All customers were entered onto the company database and with the customers permission they were kept informed of any offers in the future.

Once the costs were deducted from the takings Richard was making £3000 profit per operative per week. He had set up a team of 4 operatives in his county and after about 4 years was making £12,000 profit per week. This was just over six hundred thousand pounds a year.

His goal was to build the business to £1.2 million pounds profit. To do this he had to duplicate his system into other counties. He chose a second county, built a team of 4 and started to build the business. This time it only took just over 2 years to hit the £12,000 weekly profit goal. He started looking at a third area to expand the brand further.

Richard met with Martin and they decided that they would sell the business while they were on a roll and Martin actually had someone in mind who was prepared to pay £4 million for the business.

Two years later Richard had gotten used to the money and now £4 million did not seem that much. Ok he had to give Martin his 20% so he got £800,000 for effectively doing nothing more than give advice and a bit of money. Richard took a 6-month holiday in the sun and reflected on what had happened.

All he had really done was price on value instead of time and materials.

He wondered how he ever thought he would make any money on day rate. It all seemed so obvious to him now. He had loads of ideas too and now he had the money to carry them out. It was funny because his mates had started calling him Rich, sort of taking the mickey out of all his money.

He liked it.

A new idea was forming. He could set up a company that specialised in spraying luxury apartments. He knew of a decorating company that was charging £1000 to decorate an apartment and the decorators were managing to do 1 per week. Rich felt that if you get the systems right, he could do 3 a week, maybe even 4. A team of 4 would do 16 a week. That would be £16,000 a week turnover. This was not as good as his last business but there was much more of this type of work around the country and the volumes were higher. Rich felt that in 5 years he could build a £10 million-pound company. Oh, how his view of business had changed. He laughed at the thought of maybe offering John a job but then dismissed it just as quick. John bless him, still at the same firm they had both started at.

He got on the phone to the private training company and started discussing a training package.

Lessons learnt

Specialise in area of work so you can create your own niche. This makes your prices harder to compare with other companies.

Train your staff with the skills and the expectations of what you want from them.

Know your cost price.

Price on value and not on time taken.

Think bigger, easier said than done I know. Ok I know, I was pushing the boundaries a bit with the figures used in this story, but I have done it on purpose to try and make you think more about the business and finance side of things.

Chapter 13 – Maths – don't worry everyone is scared of this

What was the purpose of the previous chapter? Well I wanted to paint a picture of how your business could be. I know this is not what everyone wants and it's probably not why you bought the book.

If you are going to improve your pricing process, then it is going to be a journey and you need to realise how far the journey could take you if you really wanted. You are in the driving seat though so it's entirely up to you.

Now that we understand that the price is set by the market, we also understand that the market will change over time and that the market will be different in different parts of the country and different parts of the world. Because of this it is impossible to give you a set of prices that you can just take off the peg and use.

You are constantly testing the market to see what price it will bear. There are a number of ways to do this. The best way is to get feedback from customers. They may let you know when you get a job what the other prices were. They may let you know if you don't get the job too.

The only problem with this is that the prices can sometimes vary quite widely depending on who the customer has asked to quote for the job. For example, they may get a price for their lounge to be decorated. The 3 prices they get could be £150, £450, £800.

The first price may be someone who is not a decorator and is just trying it on, the second price may be a fair quote and the last price maybe someone who is so busy that they have priced the job high so that if they do get it then they can get someone on the job to do it and not lose money.

This is where you need to start doing your own sums. Before we do this, I want to talk about maths.

Maths – the very word strikes fear into your heart. How do I know this? I know this because I have been teaching apprentices and adults how to price decorating work for nearly 30 years and I know what people's reaction always is.

At first, I thought it was just young apprentices that had this fear. "I just can't do maths Pete" they tell me. Then I started teaching adults and I got the same reaction. They would stare at the floor and shuffle about a bit as I waffled on about how far paint will go. As time went by, I started to chat with people about their fear of maths before doing the lesson. They started to open up to me. I learned 2 things from these talks.

Firstly, everybody is the same, they just don't talk about it. I don't think I have had anybody say that they were confident with their ability with numbers.

Secondly, everyone blames their experience of maths when they were at school. Now I am not having a go at schools here, I think the reason is that when you do maths at school it's not really set in the real world. Once you start talking about working out real things that you understand then I think it is easier.

There is not a lot of maths involved in pricing but there is some. Lucky for us it tends to be quite repetitive and once you have done it over and over again then you will get very comfortable with it, even enjoy it. Well maybe that's just me.

Let's look at an example.

If I said that the job was £900, and the materials were £180 then could you work out what the labour cost was?

£900 - £180 = £720

Yes, you could, no problem. I find people are good at maths when it comes to money.

What about this next one?

A room is 4 metres long and 3 metres wide. How long is the coving that goes around the room?

You would just add the length of each wall.

4m + 3m + 4m + 3m = 14 metres.

This would be referred to as linear metres because you have measured this surface in a line. Narrow things are measured this way. Skirting's are another example.

I think you could do this without a problem as well.

The main point that I am trying to make is that you can easily do the level of maths that I am going to ask you to do in the next chapter, don't switch off to it. Many students just sit there with a blank look on their face because they have decided that they can't do it before they even start.

I am asking that you do not do this.

I am going to keep the sums simple to show you what you need to do, I am going to leave any overly complicated stuff out of this book.

The main objective is to get you pricing in a more effective way, once you get going you will improve your skills and gain more confidence.

When I teach decorators how to spray one of the things that interests me is why more decorators don't do it. I think it is because they are a little scared of it, I know I used to be.

I think pricing is the same thing. Deep down you have these horrible memories from school that have left you lacking confidence with numbers. Now is the time to overcome this and start again.

Just like spraying, it's worth the journey to make more money and enjoy your work more in the future.

Chapter 14 – Starting to price – a look at some of the basics

Right let's get started, I can't believe that I have got to Chapter 14 without actually talking about pricing properly. I know I do tend to go on a bit.

If you are currently pricing by guessing how long the job takes and then multiplying by the day rate, then I want you to carry on with this method for now.

This will do two things.

It will give you more confidence that the new method is actually correct, and it will just double check your maths. If for example you have guessed that you will need 5 litres of paint and your calculations say that you will need 50 litres, then you have probably hit the wrong button on the calculator.

Most trades will measure up the area to be worked on and then multiply by a rate. So, for example floor layers may have a rate of £50 a metre and they are laying flooring on 100 square metres of floor. Their price would be £5,000. This price would of course include their materials.

We need to start doing the same. It's a bit more difficult for us because we paint a whole range of surfaces and we also wallpaper. For now, we will start simple and look at working out the area of ceilings and walls.

Area of the ceiling.

To work out the area of the ceiling you multiply the length of the room by the width of the room. So, if your room is 3 metres by 4 metres then the ceiling is; -

$3m \times 4m = 12m^2$

Area of the walls

Walls are a bit different, there are 2 ways to work out the walls, an easy to understand way and a quicker way.

Easy to understand way for walls.

Our room is 3m by 4m and its 2.4m high. You need the height to work out the walls.

There are 4 walls and each one has a width and a height as follows; -

$3m \times 2.4m = 7.2m^2$

$4m \times 2.4m = 9.6m^2$

$3m \times 2.4m = 7.2m^2$

$4m \times 2.4m = 9.6m^2$

Total = 7.2 + 9.6 + 7.2 + 9.6

Total = $33.6m^2$

Quicker method for walls.

A quicker way for the walls is just to add together the wall size to give you the distance around the room (the perimeter) like this; -

3m + 4m + 3m + 4m = 14m

Then you multiply this by the height of the room; -

$14m \times 2.4m = 33.6m^2$

This is about as hard as it gets. What you need to do next is measure up your own house room by room. Write down the area of the ceiling and walls for each room. The good thing about this is that you are familiar with your own house and it will give you a feel of what different areas look like. When you measure up a real job, it's easier to know if your calculations are correct.

When you measure the walls ignore the windows and doors, except if they take up the whole walls. The time you save not having to paint that area lost by the window is offset by the time it takes to cut around it or mask it.

I bought myself a laser tape measure from Amazon for about £25. This makes the job very easy. You just put the measure on the wall, point it across the room and press a button. The display then shows the size of the room. Easier than messing about with tape measures. It's good for measuring very high ceilings too.

So far so good.

In simple terms the price is built up of two parts, the materials and the labour. However, once you have a rate then this automatically calculates the price of the paint and labour just like it did for the floor layers. You still need to know how much paint to order though. This is fairly easy to work out.

If you have a ceiling that is $12m^2$ and its smooth and previously painted, then typically 1 litre of emulsion paint will cover $12m^2$, two coats would take 2 litres so you would buy 2.5 litres.

How did I know how far the 1 litre of emulsion would go?

This information is on the back of the tin of paint and also on the data sheet. Manufacturers of paint tend to be optimistic with their numbers. For example, Dulux say that their emulsion will cover $17m^2$ per litre.

The coverage rate will change depending on how smooth or rough the surface is and also how absorbent it is. A new rendered wall that has never been painted will be both rough and very porous. It will soak up a lot of paint. This type of surface may take a whole litre just to paint $4m^2$

As a decorator you will already know how far paints go on different surfaces but it's worth keeping a little notebook and build up a set of coverage on different jobs. See how much paint that you use for each coat and write it down.

Now you can work out how much paint you need. We still do not have a price for the room though. For this I am going to give you a rate to use. This rate is for preparation and 2 coats of emulsion. We are using it just to put into our calculations. Later we will look at how to build up your own rate.

The rate we are going to use on the walls is £6 per metre. This includes paint. The ceiling was $12m^2$ so the price would be $12m^2$ X £6 = £72.

Easy.

Why would you want to build up your own rates? Well let's say that when you put in all your costs, your time, you van etc and then you add in the price that you pay for paint then you may find that your rate is £3.00 a metre.

This is your cost price.

The £6 per metre is the market price and will vary up and down as we discussed in the previous chapters. If you are pricing for a job with a builder and they come back to you and tell you that you are too expensive, and they are only prepared to pay £2.50 a metre then you know that you cannot do this because you would lose money as it is less than your £3.00 cost price.

We need to work on our cost price to try and get it as low as possible. This is for two reasons, firstly the lower that your cost price is then the bigger the difference between your cost and the market price. This means you make more profit.

Say you get your cost price down to £1.50 a metre by working faster (spraying for example) and you negotiate a deal with your supplier to get your paint cheaper then you are making; -

£6 minus £1.50 = £4.50 profit per metre.

Wow that's a lot!

Some decorators actually work more slowly on a job so that they can charge more days for the job, that's not really a good approach.

This way you have a real incentive to decrease costs and increase profit. There are pitfalls with this, but we will look at them in another chapter.

What have we learnt so far?

1. We can work out the area of a ceiling
2. We can work out the area of all the walls
3. We can work out how much paint that we need
4. We know what a rate is and how to use it.
5. We know what the cost price is but not how to work it out.

In the next chapter we are going to look at how to calculate our own rates and then price up a job.

Chapter 15 – Pricing some more

We have arrived at the point where can actually have a go at some pricing. We understand how to work out the area of a wall or ceiling and also the length of the skirting and architrave if needed. If you have been decorating for any length of time you will know how much that you pay for paint at suppliers and you can start to use these prices when building up your rate.

There are a few ways that you can go about this, but I will talk about three approaches. I will start with what I think is the best approach.

1. Working out your own rates.

To do this you need to start timing yourself on decorating jobs. I do this all the time. It's important that you write down what your times are.

I will give you a few tips on this. First of all, time the entire job. So, if you are decorating a lounge and you are emulsioning the ceiling white, the walls grey and the woodwork white then make a note of the day and time you started, log all your hours and then note the day and time you finished. So, for example it may have taken 22 hours overall.

Next, time each process. So, for example how long did you spend on prepping, filling and sanding and the like. Be honest with yourself. If you arrived at 8:00 am and you finished prepping at 12:30 pm then the work took 4.5 hours.

Don't deduct breaks and don't rush.

Time emulsioning the ceiling to a finish. Time emulsioning the walls as well and also time painting the woodwork. You can time as many processes as you like, you can even overlap some. You could time overall prepping and you could time prepping and painting the ceiling. The more you understand how long it takes you to do things the more accurate you will be when working out your base cost price.

Finally measure up the room. How many linear metres of skirting and architrave are there? How many square metres of ceiling? How many square metres of wall?

At first you will find this a bit of a chore, but it gets easier and the results are very useful, and the process is essential to building your pricing system. If you don't do this or you do it poorly then you will never fully have a handle on your pricing, even if you use someone else's rates. The longer you do this the more information you will have on how you work. Keep it all in a little black book, in your bag at all times.

Working out your cost price.

Don't forget, cost price is just what it costs us to do the job, it's not the price to the customer.

Your cost price includes; -

Your time
The materials
Overheads

First of all, your time.

You need to decide what you want to earn per year. This is not what you will actually earn it's just a figure that if you earn any less it's not worth running the business because you could make more money in a job. For example, if you earned £22,000 per year at your old decorating company then you want at least that amount in your business.

You will work about 40 hours x 48 weeks a year. So that's 1,920 hours.

£22,000 divided by 1,920 hours is **£11.45 per hour**.

Remember this we will use it later.

Your overheads.

You need to list your overheads and how much they cost. For example, your van, your accountant, your website etc. This will vary from company to company and of course you want to keep this as low as you can. I am going to assume that your overheads are £5,000 per year.

£5,000 divided by 1,920 hours is **£2.60 per hour**.

Remember this too. You can add the 2 numbers together (£11.45 + £2.60) if you like but for now, I am going to keep them separate.

Profit

Let's talk a little bit about profit. Profit is the difference between the price you charge the customer and your cost price (less materials of course).

A rule of thumb that is useful to keep in mind is if the price to the customer is £1000 then the four parts will break down as follows; -

Labour cost £500 or 50%
Material cost £200 or 20%
Overhead cost £100 or 10%
Profit £200 or 20%

Obviously, this is a massive simplification, but we need something.

If you price a job up and you think it will take 22 hours (like our lounge above) then what would be the price?

First of all, our cost price.

Labour - 22 hours X £11.45 =	£251.90
Materials =	£80.00
Overheads – 22 hours X £2.60 =	£57.20
This is a cost price of	**£389.10**

We know that we normally would charge £470 for emulsioning a lounge. This is our market price.

£470 - £389.10 = **£80.90 profit.**

Right now, let's have a final look at our figures.

The job was priced at £470 so our rule of thumb would give; -

Labour =	£235 (50%) –	was actually £251.90
Materials =	£94 (20%) –	was actually £80.00
Overheads =	£47 (10%) –	was actually £57.20
Profit =	£94 (20%) –	was actually £80.90

Hopefully you can see that the rule of thumb is just a rough guide, but you can also see that it actually pans out quite accurately.

You can also see that we are a bit high on labour and a bit low on profit.

This is a great start to our understanding of how we price and it's an important first step to take. Keep a record of all your prices and break them down like this.

I also keep up to date with what are the typical prices that are being charged by other decorators. Not so that I can copy them but so that I understand the market.

For example, if my mate Bob normally charges £400 for a lounge but this year, he is charging £480 (and getting the work) then I know that prices are on the rise and I will adjust my prices accordingly.

A lot of this looks like day rate, what was all that about areas you were going on about before?

Converting your prices to an area rate.

Ok so this is where we start to build our own rates.

Let's say the ceiling was $20m^{2,}$
It took me 4 hours to prep and paint,
It took 3.3 litres to paint (say a 5 litre tin).

Here is the calculation for cost price for the ceiling; -

4hours X £11.45 (labour) =	£45.80
4 hours X £2.60 (overheads) =	£10.40
Materials =	£30
Total =	**£86.20**

A celling with an area of $20m^2$ **COSTS** us £86.20 to paint.

That's £86.20 divided by $20m^2$ = **£4.31 per metre.**

Decorators that I know will price between £4.00 and £8.00 per metre to prepare and emulsion a ceiling with 2 coats.

You know that you can't go any lower that £4.31 but if you went in at £6.00 you would make £1.69 profit per metre which is more than 20% so happy days.

In future to price a ceiling you could just multiply the area by your rate of £6.00.

For example, $20m^2 \times £6.00 = £120.00$.

The main thing is that you understand why you are charging that and monitor it from job to job. You can also see what rates are being used by other decorators and break it down to see if it would work for you as we did with the ceiling rate above.

Linear metres.

Skirting and architraves are slightly different in that we charge by linear metre and not area. This just means that you multiply your rate by the length of the skirting.

If the skirting is really deep, then you could revert back to area. The standard practice is that anything deeper than 300mm is calculated by area. The area would just be the length of the skirting multiplied by the depth.

Try and keep your rates so that they include your time, your overheads and the materials. If you are using a very expensive material, then you could add extra onto your rate to compensate for this or just add the price of the material on at the end.

Here is the calculation for cost price for the skirting: -

3 hours X £11.45 (labour) =	£34.35
3 hours X £2.60 (overheads) =	£7.80
Materials =	£10
Total =	**£52.15**

The room is 5m X 4m so that you have 18 m of skirting round the room.

£52.15 divided by 18m = £2.89 per linear metre.

This means you can't charge less than £2.89 or you are losing money.

If you charged £3.60 per linear metre, then you would earn: -

£3.60 X 18m = £64.80

£64.80 (our actual price),
minus £52.15 (cost price),
equals £12.65 (profit).

There is a temptation at this point for you to just use my number of £3.60 per metre when you price skirting boards and architraves but I don't want you to really do this.

You may find that you are brilliant at woodwork and your gloss work is amazing and because of that your customers are willing to pay £4.00 a metre. If that is the case, then you should charge that.

The main thing is that you understand your cost price and your actual price and your profit.

What about wallpaper?

Wallpaper is straight forward to price for. You can do it a couple of ways,

1. Price per roll.
2. Price per metre.

I will do an example for both.

Here is the calculation for cost price for wallpaper per roll. 1 roll would cost.

1 hour X £11.45 (labour) =	£11.45
1 hour X £2.60 (overheads) =	£2.60
Materials (cost of the roll) =	£50
Total =	**£64.05**

On this job your COST price per roll would be £64.05.

You could exclude the wallpaper and just add that onto the price.

Here is the calculation for cost price for wallpaper per roll. 1 roll would cost (excluding materials).

1 hour X £11.45 (labour) =	£11.45
1 hour X £2.60 (overheads) =	£2.60
Total =	**£14.05**

I have asked decorators what they charge per roll and its anything between £25 and £40 per roll (excluding wallpaper) so you can see a healthy profit margin there.

Of course, if you are hanging a very expensive wallpaper you would want more because there is more liability. More things could go wrong and cost you money to replace the paper.

Also, I would have a separate rate for stripping the walls and preparing them for painting, the above rate is just for hanging the wallpaper.

To calculate the cost per metre you would just take your cost per roll, in our case £14.05 and divide it by 4.5m^2 which is what a standard roll of wallpaper covers. Look on the label for this information. I have deducted 10% for wastage.

£14.05 divided by 4.5m = £3.12 per metre.

This is our COST price per metre. Most decorators will charge between £5.50 and £10.00 a metre.

2. Using a pricing book to get your rates.

If working out your own rates seems like a lot of work, then another approach is to use a pricing book such as SPONS. Look it up on Amazon, they bring a new one out every year to keep pace with price changes but are quite expensive (about £150) most people buy one and then use it for a number of years and add a percentage to the rates to account for inflation.

There are advantages and disadvantages to using a pricing book. The advantages are that all the work has been done for you and you can see how they worked it out at the beginning of the book. This information can be useful if you are learning about pricing and you want some guidance.

Another advantage is that the book lists every conceivable situation that you may come across and gives you a rate. Sometimes it can make a good starting point for your own pricing.

The disadvantages apart from the high cost of the book is that it can be quite mind blowing when you first look through it. It seems quite complicated. Also, it's not tailored to your own business and is just showing industry standard prices which, you may want to get away from.

If you are serious about pricing though it's worth getting a copy to look at even if it's a second hand one. The Kindle version is quite a bit cheaper too, and you could view that on your computer.

3. Using a professional

The final approach may be a good idea if you are just starting out in business or just getting into pricing using rates. You could spend some time with a professional estimator, ideally a local one and one who deals with decorating work. They will work with you to price jobs and help you understand how to build your own rates.

If you have done as I suggested and worked out your cost price for labour and overheads and also timed yourself on some of your jobs, then the estimator will be able to help you even better.

If you get a big job, then the estimator can just price the job for you for a fee and if it's a smaller job they could just guide you for an hourly rate. Well worth finding someone to help you on your journey.

Chapter 16 - The Art of Estimating Paperless

This chapter has been written by Chris Berry – "The Idaho Painter" to give an insight into his approach to pricing. I think that a lot of people over complicate the pricing process and I feel that it could be simplified, Chris uses a system that allows him to price without even going to the job and make a good profit as well.

I will hand you over to him.....

Having been the owner of a highly successful painting company in Boise Idaho for 18 years, I had the opportunity to perfect my estimates through writing thousands of them. Estimating was done in the traditional manner in the beginning of my career by meeting with a customer, gathering details in person as I walked the job with the client, traveling back to the office to write the bid, printing and compiling the document in a custom folder, then hand delivering the estimate to the customer in a large bid pack tailored for them.

With the rise of the internet and power of social media the bidding process has drastically changed over the years to a simpler process. If you stay with me here, I am going to explain how you can write a bid and complete the job without ever meeting the customer.

In the beginning you will want to meet every potential client to sell yourself and your company. But as your business grows and if you are doing things right the referrals should start to roll in and this is when you can begin to incorporate my unusual bidding process. Our old process included our printed copy of our 20-page bid package.

This package is essential to selling your company and positioning you as the most qualified candidate for the job. The majority of your potential clients will be seeking other bids to compare rates as well as who they think will be the best fit for their job. Once they have obtained multiple bids they will be sitting down and evaluating them to determine who to hire, this is your opportunity to sell yourself.

If you have a 20-page document that answers all their potential questions and outlines exactly what you are going to do for them, you are creating a large advantage in your favour. From my experience a vast majority of painters have handed in a 1-page hand scratched bid or a carbon copy 2 page bid.

With the 3 bids sitting there who do you think has the best odds of winning the bid? Over the years our company became so highly referred because of our work ethic, customer service, and quality paint jobs that the majority of referrals did not get comparison bids. This takes time but this is what you are after and can be achieved.

As I began to build my reputation the referrals began pouring in and so did my lack of time to write them. I was so swamped painting all day with my crew then meeting customers at the end of the day for bids and colour consults that I decided I had to come up with a more efficient bidding process.

This is where the power of the internet comes into play. I quickly discovered that Google Maps is very powerful and has photographs from a street view that is almost as good as being at the house live. Well maybe not that good but good enough to write an exterior bid from.

Now you are probably thinking, what about the back of the house or the smaller details not included in the picture. This is where another app fills in the blanks. Zillow will pull up all the information on a home including the important square footage details that I am after when writing a bid.

This real estate app will list more information than you will possibly need. You can look at the view of the house from Google Maps that is like a 3D walk of the front and you can position yourself wherever you want on the street so you can quickly study the house for details that will give you an insight on how difficult it will be to paint. You are able to gather information like, will you need a lift, how many potential colours it will be, prep work involved and so forth. Zillow adds details like the age of the home which can give insight on whether there is lead in the paint, how much prep work due to the age of the home, how much paint will be involved using the square feet listed etc....

Are you still a bit sceptical about bidding a place you did not physically see in person? This is where experience comes into play. Google Maps and Google Earth will give you more details on whether the house is potentially one story up front and 3 stories in back. Yes, even with all the details you gather you can still have that unusual circumstance that could not be seen.

But this is where the law of averages works out. We were painting 3-5 houses a week and in one circumstance we did run into the house that was one story up front and 3 in the back but the same month we had circumstance just the opposite that made the job a little more profitable.

Over a span of 10 years bidding in this manner it always averaged out for us in our favour. Now if this terrifies you, you can still bid in this manner but just do a quick drive by of the home. A drive by may build your confidence and that still saves potential hours not having to meet with the customer. Meeting the customers, in many cases leads to them wanting colour advice and other consulting conversations before you even have a contract which cuts into time you could be painting and making money.

But hear me out here, I did not start out this way. I built my company into this simple and effective bidding process by putting business practices into place day one when I started B&K Painting. Those practices were building a company on strong ethics, values, quality employees, client respect, and being thankful for each and every client who was helping pay my bills.

Using Google Maps, Google Earth, and Zillow I could gather just about everything I needed to write an effective exterior bid which was the primary focus of my business model, doing exterior residential repaints. Zillow and many real-estate apps have video walk throughs of homes now a days. These are advantages that I used, and they worked out in my favour. I can now sit at my desk and get a walkthrough of the home where someone has flown a drone on the outside and walked through the inside giving you a 3D view of the home.

Crazy the thought that someone can look inside your home, everything about your home is public information. It is quite incredible. Now sitting at my desk and on my iMac I gather all this information and come up with a price for the bid. Our prices for an exterior repaint are based on an exterior square foot price. That is the exterior square foot (footprint) of the house. Our square foot prices rise and fall according to many factors including, demand (how booked we are), age of the home, how many stories, how much prep is foreseen etc. There is a human element that is added to our price along with experience and discretion. If I get a feeling I need to go look the house that is what I do.

A common average price as of writing this article is $1.75 a square foot in Boise Idaho. I simply take $1.75 and multiply that by the square foot of the house footprint. If it is 3500 square feet, then the math is simple. $1.75 x 3500 = $6125.00 That is the price I write in the estimate. That number is then entered into Quickbooks in the estimate. A one page estimate is created as a PDF and emailed to the client.

Quickbooks has the ability to have custom Word docs entered and a few clicks of a button and a 20 page bid package with the clients' name throughout is generated and emailed to them. And there you have it, saving trees, ink, time, and money and the customer get a personalized bid pack and estimate.

Now of course you are going ask, what about interiors. Yes, our method works great for exteriors, but interiors do pose another challenge. You cannot do a drive by. But remember many homes have video walk throughs on real-estate apps. Many times, this along with the data you gather from Zillow will be sufficient. With that said I would say we have to meet with clients for about 50% of interior bids.

Cabinets on the other hand are very simple. We charge per door and drawer front with only a few other details that may increase the bid. As of writing this article we bid cabinets at $135 per door and drawer front. Added charges include oak grain cabinets, doors with glass, multiple colours, and filling knots in cabinets with wood like knotty alder.

Most cabinet bids can be written by asking your potential client how many doors and drawers and what type of wood the cabinets are made of. We do ask the clients to send us pictures of the cabinets if we are comfortable writing a bid without a visit. About 90% of the time this will be enough information for me to write a bid. Quick, efficient, and saves times.

Once again, this process works when you are confident you do not have to meet the customer in person to sell yourself or you are getting so many bid requests you can write more bids and get less of the jobs, but the averages work out. This method also takes into account that you know your crew well and the output they can generate.

On exteriors any house under 3500 square feet our crew could paint in 1 day no matter how many colours.

We had a fast and efficient crew, think that's impossible? Now of course I am talking about painting, not pressure washing and prepping which happens on another day.

In the beginning of my career I was meeting every potential client and had a win percentage of around 75%. With our streamlined bidding process, it dropped to around 45% but we were taking and writing 10 times more bids, so the numbers continued to work out. Our painting company typically stayed booked out for about 3 months and if we ever fell to the one month area, I would simply start meeting with potential clients again.

I would meet clients until we created enough demand that our prices would go back up and we were booked out for 3 months. The farther out we got booked the higher our prices got. Very simple law of supply and demand. So now I will give you the fly on the wall look of how all this happened.

Typically, a call would come into the office and my office assistant otherwise known as my wife, Lisa would answer the phone in a very professional manner. Customer says they would like to get an estimate. Lisa would ask how they heard of us. If it is a referral Lisa gathers the referral information to send a gift card to the referrer. Lisa then gathers information about the job that could help in writing a bid without going to the site.

If it is fence, how many fence panels? How many cabinet doors? Address, Email, phone number, age of the home, how soon do you need it done? If they need it tomorrow and you are booked for 3 months you are wasting valuable time writing a bid. Interior or exterior? Gathering information on this phone call is vital. The more information you gather the better odds you will be able to write a good and accurate estimate. Lisa then informs the caller that a bid will be written, and an email sent with the bid attached for their review.

During the conversation she informs them that if they are comfortable with the estimate price but want to meet with me (Chris Berry) she will schedule an appointment to meet and go over the Job. About 10% of clients request this meeting. Many accept the job, email the garage code and we go to work without even meeting them. This is what happens when you build a business built on trust.

You have to be willing to run a company that changes that stereotype of contractors. Be professional, look professional, wear whites, be polite and courteous to name a few. Do not roll up with offensive music playing, using bad language, smelling like alcohol, get to work right away, cover offensive ink, be on time every time to name a few more.

Our model has worked well for our painting company that lasted 18 years and eventually sold as a very successful and highly recommended business in Boise Idaho. Our model was also based on good estimating practices that included a good profit margin, wise estimating practices, having a good understanding of material cost of each job, knowing your companies daily operating cost, and always adding a decent percentage of the bid as profit.

If you are interested in seeing my bid package that has won thousands of bids over the years without even meeting a customer, it is available on our website www.idahopainter.com simply click the link to visit the page. My bid package was not written overnight, it has been constantly evolving and has been in the works for about eight years. Every year I add and make changes according to circumstances that arise throughout the year.

Chapter 17 – Pricing from plans

I have left this one until near the end and in fairness you could probably dedicate a whole book to this topic as well. When you first start decorating for yourself you will tend to do work for family and friends.

The work that you do will be existing properties that you can go and have a look at. This of course makes it easier to estimate how long it will take you because you can visualise it. It also makes it easier to spot any time consuming elements.

If you start doing commercial work or even larger domestic jobs that involve extensions, then you will be expected at some point to produce a price from a plan.

Many of the other trades are used to working from plans however we tend not to, and by the time we come along the building is almost complete. For this reason, when a decorator first sees a set of plans, they are a little daunted.

Don’t be.

The good news about pricing from plans is; -

1. It's not as hard as it looks.
2. You can pay someone to do it if you like.

My advice would be to get hold of a set of plans for something fairly simple like a house extension, ideally one of a friend or neighbours where you can go and view the completed build and compare it to the plans.

That way you can look at the actual extension and price it like you would normally and then you could go away and price it off the plan and see how close the two numbers are. This way you will build up confidence of your ability to understand plans and also measure off them.

Another approach would be to pay someone to price off some plans for you if you get the opportunity to do this. Then once you get the job and do it you can refer back to the plans and compare it with the real building.

One of the problems with working from plans is that it can be quite time consuming if you are not familiar with them and you can spend all day looking at them and measuring up only to find once you have put the price in that you did not get the job. You will find though that the more that you do it the better you get and the easier it becomes.

Scale

You cannot draw the proposed building or extension full size, well you could I suppose but it would be a massive piece of paper and would not fit on your dining table. Because of this, the drawings are done to scale. This means that they are so many times smaller than the real thing. For example, something that has been drawn 1:50 is fifty times smaller than the real building will be.

The bigger the real item is going to be the bigger the scale. If you have the drawing of the whole housing estate, then it may be 2500 times smaller than the real thing. There are a handful of common scales that are used in the construction industry.

If you have a drawing that you are working from then it is handy to have a scale rule so that you can measure off the drawing. This is not needed for everything because measurements will be marked on the drawing however it is handy for certain overall measurements or for smaller items that have no measurement on them.

Take off your measurements.

Once you get your set of drawings then you need to measure each room just like you would in real life and work out the ceiling and wall areas. You also need to measure skirting lengths, architraves, doors and windows. Plus of course any other elements that are being painted.

Once you have these sizes you can price them just like you would if you had got the sizes from a real building by multiplying your areas by your rates that you have built up.

You can now see one of the big advantages of having rates for both square metres and linear metres because they make pricing from plans so much easier than if you are used to guessing how long things take.

A worked example

I have given this much thought and it would be difficult to include full size plans in this little book. If you want a set of plans and a worked example for working out the areas, then email me on; -

fastandflawless@outlook.co.uk

I will post a pack out to you. There will be a small cost to cover printing and postage.

I am also looking to write some additional chapters to add to the book in future editions and when I do then I will email these out to you for free.

Don't worry I will not spam you, it is not my style.

Chapter 18 - Handling customers

Decorators spend a lot of time thinking about products, pricing and how to do the job correctly but I don't think many of us think too much about the types of customers we are going to work for and how to handle them so that the process of going into their home and decorating is painless for both ourselves and the customer.

Some customers can be lovely, and some can be hell. Most are somewhere in between and it's a good idea to polish your people skills at this point because it could lead to more work.

Before we start - Presenting the right image

The subject of wearing white overalls is much discussed amongst decorators (for some reason) and I am going to throw my opinion in here for what it's worth.

To me overalls are like a uniform so that people know what job you do and that you take that role seriously. It's like a policeman or a nurse or an airline pilot, I could go on, there are loads.

Imagine getting on a plane and the pilot was wearing jogging pants and a string vest, would you be confident he was a “good” pilot. What if he assured you that he was brilliant, would you believe him?

What about a policeman (or woman for that matter) imagine if you were cut up on the road and pulled over by an ordinary car and the person that got out was scruffy and unshaven? No uniform in sight.

Would you even believe they were the police?

Well we are the same as decorators. If you turn up to the customers house on the first day and you are smart, clean and wearing overalls then the customer makes a subconscious judgement that you are competent and professional.

All you have to do is go and prove them right.

If you turn up scruffy in your torn jeans and a t shirt then believe me, they are going to look very closely at your work when you leave. That’s if they even let you start.

The same goes for your van. Let’s have it clean and well-presented and if you have decided to get it branded up then make sure that it is professionally done. That way your customers first impression is what you want it to be.

New customers – check they are right for you

Many years ago, when my kids were in primary school one of the mothers approached me to do some decorating.

"I just need my lounge decorating" she said.

No problem, I can come up and have a look and then give you a price. Then I had a thought, did she really have an idea how much the decorating was going to cost or was she expecting "mates rates" because our kids went to the same school.

So, I added.... "roughly how much were you expecting the decorating would be?"

£100 was the deluded reply.

At this point I realised that this was not a customer I wanted, obviously I could not do the decorating for such a low price, it was not even worth my time going to look at it to price.

I politely declined the offer of work from this customer with some flimsy excuse about being really busy at the moment and also that the price would be more than likely double or treble what she was thinking, and I was off.

I always think that your ultimate mission as a self-employed decorator or owner of a decorating business is to build yourself a portfolio of really good customers. What makes a good customer is up to you. You may decide that you like working for customers that give you coffee and cake at brew time or are customers that are not too fussy.

I like my customers to be prepared to pay my price and pay on time. It's a bonus if they are quite nice people too but that's not essential because I can handle them if they are not.

So, once you know what your ideal customer then try and suss them out as early as possible to save you wasting your valuable time.

Types of customers

We don't really think about different types of customers when we first start out as a decorator with our own business, but once you get going you realise that there are indeed different types of customers. Some are good, some are hard work, and some are to be avoided at all costs because they will take you to the cleaners.

I will tell a little story to illustrate the last type.

I was subcontracting for a local company. They had asked me to spray the outside of a big new extension on a million-pound house. When I turned up I was met with a scruffy guy in overalls who said "I am Frank..... and you are?"

He went on to explain that he had had a new rubber roof installed on the extension costing thousands and if any overspray got on it then I would have to pay for a new one.

"What guarantee can you give me that no overspray will go on the roof?" he asked.

At that point I didn't say much but decided a few things.

1. He was an arsehole
2. I was not going to spray his outside
3. He would find overspray even if there was none.

I phoned the guy I was working for, let's call him Rob because that was his name and explained what had happened. "Sorry Rob, I can't work for this guy" and off I went.

Rob rolled the outside and guess what? The owner (yes that's who it was) claimed that he found overspray on the roof (even though it was rolled) and would not pay.

You see – avoid this type of customer at all costs.

Other types of customers include; -

1. Cheapskates
2. Unreasonable
3. Manipulative

I will deal with each type in turn.

Cheapskate customers

This took me a while to work out but once I did, I changed the way that I took on new customers immediately. Let me explain what happens, it's an easy trap to fall into. You get a phone call from a new customer, maybe it's someone that you know or a friend of a friend. You go and look at the job and decide that because you know them that you will give them a good price.

You say to them, "this is a good price because you are a friend but don't tell everyone that's what I have charged you or they will all want it for the same."

"We won't."

Famous last words.

So, you do the job for this customer. Let's call the customer Mrs Smith. A week later you get a phone call for some decorating work. Where did you get my number (always do this by the way) Oh Mrs Smith recommended you.

You go and look at the job, you give them the price, they haggle and sulk wanting you to drop the price and finally when you do the work it's not really worth your while. Over time you have a raft of these people who have all come from the original Mrs Smith job.

What I realised is that all these customers were cheapskates, they didn't want to pay the going rate they just wanted a cheap job. If you think about it is obvious.

Why did they call you in the first place? Mrs Smith told them that you were cheap. That's why they want you. Not because you are good at your job, or honest or clean and tidy or reliable. In a way it's your fault.

What did I do?

I binned off all the Mrs Smith customers and started again. These days if I get a sniff that the customer is "cheapskate" I tell them I am too busy and walk away.

Unreasonable customers

Some customers are just hard work. They could be called awkward. You know the type, they can't decide on the colours they want and mess you about while they decide. They pick a colour but once you put it on, they change their mind. "Are you sure that was the colour I chose?" They don't move any furniture, they make you wait for your money.

I could go on.

Some people are just impossible to deal with and seem to have their own view of the world and won't meet you halfway on anything. It has nothing to do with money, they can be rich or poor.

The best way to handle these customers (and all customers really) is to make clear at the outset what your expectations are. Some decorators call this their "terms and conditions." It's even better if you have these in writing. What this does is pre-empt any future problems.

For example, you could have it in your terms and conditions that the customer has to decide their colours a week before you start and test the colour with a “match pot” before you start. If you can’t start because they have not done this, you could defer the job.

There are many things that can cause the customer to become unreasonable and it’s up to you to cover all the bases.

Another example is what does the price include and what does it not include. We can move your furniture for you, that’s an extra £70. The Idaho Painter – Chris Berry has a “Bid package” that includes what exactly the price covers and what is extra. He makes quite a bit of money on the extras that the customer goes for.

You need to learn how to handle these kinds of customers. The upside is that once you nail it then a normal person is a breeze.

Manipulative customers

What do I mean by this? Well some people are very manipulative, if they are not very good at it then it’s obvious, if they are brilliant at it then you don’t even notice, and they will have you doing what they want without you even realising.

I have been a teacher for most of my life and I know most techniques out there and some are quite powerful. It's worth getting a book on it and getting up to speed yourself, that way you can counter their manipulation.

I can give you an example of pretty low level manipulation. I went to price an outside, this was years ago. The guy was a salesman and by the look of the house a pretty successful one. I walked round the house and gave him a price. I can't remember how much it was but for the sake of argument let's say it was £1800

"£1800! Did you hear that Doris? (his wife) it is £1800 for the outside of the house to be painted."

"Could you not do me a better deal? I may have a word with our window cleaner and see if he could do it."

Yes, really, he said window cleaner.

"That's no problem" I said "I am really busy, I will let you get back to me. Don't wait too long though because I may not be able to fit you in."

This really left him nowhere to go and he gave me the job and in fairness turned out to be a good customer and paid on time.

The main thing is to be aware that customers can manipulate you and develop some counter strategies. Have a look on Amazon for a good book or just Google "Manipulation techniques" if you are not even aware that this goes on then you will be amazed.

I try and avoid using manipulation techniques on people unless they do it to me first, I prefer an honest and transparent approach in all dealings with customers, it makes for a much happier life.

Builders – their own category!

I have saved the best until last. If you have had dealings with builders, then this will make you smile in recognition of what they are like. If you have not had any dealings, then this could be useful.

Before I start let's look at the possible advantages of working with a builder, you could have a steady stream of work lasting for months or even years, you could be dealing with just one person most of the time and get regular payment.

You could make some good money if you are a productive decorator. There are some good builders out there and if you can build a relationship with a good one it can be great for both parties.

Builders can be all three of our customer types rolled into one.

They can be cheapskate, they want discount for volume of course! Some of the prices out there are very low.

They can be unreasonable, they want all 10 houses finished by the end of the month, people are moving in!!

They can be manipulative. They will use power plays, bullying and try and belittle you just to get their own way.

For this reason, many decorators stay away. But learn how to handle them and you could be onto a winner.

Chapter 19 – Deposits

Whenever I speak to decorators about pricing, I always bring up the topic of deposits, it has a mixed reaction, and everyone has their own take on them.

First of all, most decorators will go and do an estimate for free and leave the customer an estimate. Then if they get the job, they will do it and cover all the cost of materials and labour themselves and then finally once complete will wait for a month or two while the customer decides to pay.

Does this sound a fair system to you? Why do we have to bankroll the project for the customer? Why are we buying all their materials for them? Finally, why are we shouldering all the risk while they take none?

I have no idea, I think maybe we have some idea that the customer is doing us a massive favour by giving us work and the most we can do is repay them by funding the work.

Wait a minute they are not doing us a favour here. Granted the customer is important and if we want to build a successful business we need to look after our customers, but the customer is getting a pretty good deal with us too. Good decorators are hard to find, and the customer is pretty lucky to have found you, so lucky in fact that they are going to meet you halfway with a deposit.

It's up to you how you work the deposit and how much you charge, I will discuss some different deposit systems that people I know use.

If you are a bit shy then maybe just charge 20% of the total price of the job payable on the day that you start. This will at least cover materials and means that you are only giving them your time on credit. On a £1000 job you would be asking for a £200 deposit, I think that is pretty fair.

One decorator I know takes 20% deposit when the customer books the job in, this could be months before they start. 30% on the first day of the job and then 50% on completion.

Another takes 50% when the job is booked in and 50% on completion, this way the risk of the job is split half and half between the contractor and the customer.

I could go on. What I like about a deposit is that it's a test to see what the customer is like. If they feel that a deposit is unreasonable and don't want to pay it then it could mean a number of things.

It could mean that they don't trust you, if this is the case then do you really want to work with them, or do you need to work a bit harder so that they do trust you. It could mean that they have no intention of paying you at all so even 20% is a bit much. Obviously run a mile if you think this is the case.

Every single customer I have asked for a deposit have paid it no problem and not batted an eyelid or commented. They know that it's just the way that I work.

Some would say that it's not professional and that established companies just do not ask for deposits. This is nonsense in my view and if you look across many industries it's quite normal to ask for a deposit. I have just bought a new car from Ford and they asked for a deposit, they are a pretty established dealership and are by no means unprofessional.

Some say that it's just not normal practice. Well I would say that we need to change the practice so that it is normal, in most cases the larger companies are just bullying you.

Don't let them.

Larger companies in general will say that "it's not our policy" but in my experience if you tell them that it is your policy and you speak to someone high enough they will make an exception for you, especially if you are brilliant at what you do and you are the leading contractor in your selected niche. At the end of the day a large company can be "an all your eggs in one basket" proposition and if you are going down that road with them then you want the deal to be a good one.

My advice would be to charge a deposit and get comfortable with it, it makes your cash flow a little bit easier.

Stage payments are another idea and you would be amazed how long some companies will make you wait for your money, personally if it's a big job I would want money every week and if they are not happy with that then I would walk away.

Some decorators break the job up into weekly chunks, for example if they are decorating the whole house, they will invoice either a room at a time or a stage at a time. As long as you are clear at the outset there is no problem and if they don't pay at the end of the first stage then alarm bells should be ringing.

Chapter 20 – Different pricing strategies

I have included this chapter just for a bit of fun really but also to make you think about your pricing in a unique way. You don't always have to follow the crowd, especially if the crowd appear broke. It is your business and you can run it any way that you please.

The following are a number of pricing strategies that other companies use when selling their wares.

Fixed price

For example, a mars bar or a washing machine. You go into the shop, the price is on the item, you pay and then take the item away. Pretty standard. It's hard for us to do this but not impossible if you do a specific type of work. For example, if you spray shop shutters you could have a menu of prices depending on the size of the shop.

Bespoke price for each customer

This is what most trades tend to do, they look at the job and give the customer a price for that specific job. Not transferrable to another job or another time. In contrast to the mars bar example where you would expect the price to be the same next time you bought one.

Monthly Subscription

This is quite a common pricing model these days and has a lot of benefits, common business that use this model are Netflix, Amazon prime, Sky, magazines etc.

I am not sure if we could use this model but a subscription fee to cover a maintenance of rental properties could work. The advantages of a subscription are that if it is paid monthly then it gives you a fairly predictable income stream.

Percentage of value

I know a decorator that sprays kitchens and he charge 10% of the value of the kitchen. So that for example if it is a £40,000 kitchen then he would charge £4,000 to spray it. Regardless of time taken. It usually takes 3 to 4 days.

You could apply this to other situations. For example, if you are wallpapering then you could charge 50% the cost of the wallpaper per roll. I think this one could work for us in certain circumstances, you just need to get creative.

Cost plus percentage

This is similar to what I have been talking about in this book. You work out your cost and then add a percentage profit. Very workable for a decorator but I think could sometimes leave some money on the table.

Day rates

We know what these are. Try and avoid like the plague.

Hourly rates

Great for employed people but not great for us.

Industry standard (SPONS)

This is good to be aware of so that we know where our prices are in relation to the industry standard. Better to try and separate yourself from the industry a bit so that it's difficult for companies to play the industry standard card.

For example, if you are a specialist in door refinishing and the builder tries to play the "my last painter charged this" card, you can reply with "I am a specialist and price accordingly."

Piecework (so much per door)

When I price to paint a door, I usually charge £10 per side per coat. So, a door that has 2 coats on both sides would be £40. This is an example of piecework. You get so much per door and the more that you do the more you are paid. It could work well if you do a lot of bespoke joinery painting work and they make a lot of the same thing and want a piecework price off you.

Annual salary

Again, good for employed people not great if you actually want to make some money. Possible if you do some specialist work for one company that is hard to quantify (training for example) and you work for them on a yearly or monthly fee.

Base price plus optional extras

Some decorators will do this when they price a job. They will give the customer a base price and then have "optional extras" that the customer can choose from.

For example

Room clearance - £70
More than 2 colours - £80 per extra colour
More than 2 coats of paint - £50 per extra coat
Colour changes - £100 per change
Extra durable paint - £150

I am not sure if these are great examples, email me if you can think if some good ones, this is a great way to upsell what you do.

Seasonal pricing

Many decorators do this too. They may charge more in summer when everyone is busy and less in January when there is less work about. I know some that will vary their rate depending on how busy they are.

If they are booked up for 3 months, then they increase their rate by 20%. If they are only booked 1 week in advance, then they may drop their rate by 20%.

Personally, I am not a big fan of this approach, but it does have its merits. It's our version of having a sale.

Two for one pricing

I sometimes do this, well not quite two for one but I will do two rooms for less if they are done together rather than doing one room now and one room in three months' time.

Price matching

Again, personally I am not a big fan, it always seems to end up in a race to the bottom however some do use this approach, especially if they cannot price. They will get the customer to get a price from another decorator and then do if for £100 cheaper. This is not a good long-term strategy but could be done in certain circumstances.

Believe it or not these are just a handful out of many prices strategies and just goes to show that there is more than one way to go about it.

Chapter 21 – A few parting words

I hope that you have found this book useful. I wanted to break the "day rate" mould that everyone appears to be using but at the same time explore possibilities for your decorating business.

I hope that you enjoyed the little story that I have weaved into the book. I thought it was a good way to get an insight into a decorating business at all levels. If you are an apprentice then much of Richards story will be new to you, if you have been decorating most of your life then I am sure that a lot of the things that happened to Richard on his journey will have made you smile.

I have pushed the boundaries with the story, and I know that many of us do not know many millionaire decorators, I actually do know a few so it's been a bit easier for me to build it into the story.

I think it's important to raise our game as decorators and I think this means doing a few things. We need to increase our skill level so that we are turning out work that your man on the street cannot achieve.

We also need to take seriously what we are doing and be proud of what we offer. We need to keep working on our skills to stay current and we need to keep up to date with products.

It's amazing how many painters or decorators don't know much about paint and will rely on their supplier to get the knowledge that they need. If we were talking to a Joiner and he knew nothing about wood I don't think we would be very impressed.

Finally, on the pricing front, I know how difficult it is to price decorating work because I have struggled with it myself. It's worth getting good at and worth the effort to build your own pricing system.

Once you have a set of rates that you know work then you will become more confident when pricing, quicker when pricing and also you will make better money.

If you have any feedback on this book or you just want to send me your thoughts then email me at fastandflawless@outlook.co.uk you never know you may get a mention in the next edition of this book.

If you have not read my previous book "Fast and Flawless – a guide to airless spraying" then check it out on Amazon and get yourself a copy.

Finally, I if you have enjoyed this book or my previous book please leave a review on Amazon. I would really appreciate it.

About the author.

Pete Wilkinson has been decorating all of his life. He currently teaches spraying courses at PaintTech Training Academy and does some select spraying contract work. He lives in Preston with his wife Tracey and in the very rare time off likes to relax on his boat.

Printed in Great Britain
by Amazon

34574452R00106